# How To Rebuild FORD Power Stroke DIESEL Engines 1994-2007

Bob McDonald

CarTech®

**CarTech®**

CarTech®, Inc.
6118 Main Street
North Branch, MN 55056
Phone: 651-277-1200 or 800-551-4754
Fax: 651-277-1203
www.cartechbooks.com



Edit by Scott Parkhurst
Layout by Monica Seiberlich

ISBN 978-1-934709-61-0
Item No. SA213

Library of Congress Cataloging-in-Publication Data

McDonald, Bob
How to rebuild Ford power stroke diesel engines 1994-2007 / by Bob McDonald.
p. cm.
ISBN 978-1-934709-61-0
1. Diesel motor--Maintenance and repair. 2. Ford automobile--Motor (Diesel)--Design and construction. I. Title.

TJ799.M33 2012
629.25'060288--dc23

2012002711

Written, edited, and designed in the U.S.A.
Printed in China.
10 9 8



***Title Page:***
*The best piston ring seal comes from the way the block is honed. Piston ring manufacturers often advise the proper bore finish for their piston rings.*

***Back Cover Photos***

***Top Left:***
*Using an 8-mm socket, remove the bolt in the center of the rocker arms below the injector. This bolt holds the injector to the cylinder head.*

***Top Right:***
*Lubricate the wrist pins with oil and assemble the pistons to the rods. Make sure to place the piston bowl toward the cam side for the 7.3.*

***Middle Left:***
*Install the piston rings onto the piston and lightly oil them. Place the rings in the proper position on the piston and install the pistons in the bore with a piston ring squeezer.*

***Middle Right:***
*Install the HPOP in the back of the front cover while positioning the gear over the shaft of the pump.*

***Bottom Left:***
*The rear seal needs to be installed. This requires a special tool in order to install it with a wear sleeve. The tool can probably be leased from your nearest parts store. If you choose to purchase the tool from OTC, the price is around $700.*

***Bottom Right:***
*Install the new cup onto the driver tool. The driver tool is machined just like a real injector. This way the nozzle in the end of the driver is used to align the cup in the cylinder head.*

DISTRIBUTION BY:

*Europe*
PGUK
63 Hatton Garden
London EC1N 8LE, England
Phone: 020 7061 1980 • Fax: 020 7242 3725
www.pguk.co.uk

*Australia*
Renniks Publications Ltd.
3/37-39 Green Street
Banksmeadow, NSW 2109, Australia
Phone: 2 9695 7055 • Fax: 2 9695 7355
www.renniks.com

*Canada*
Login Canada
300 Saulteaux Crescent
Winnipeg, MB, R3J 3T2 Canada
Phone: 800 665 1148 • Fax: 800 665 0103
www.lb.ca

# CONTENTS

# PREFACE

The intrigue and desire to write this book started from frustration at the lack of books published about Diesel engines. I often found myself at the local bookstore wanting to research a Diesel subject only to walk away empty handed. Having a desire for Diesels only propels me to learn more as technological advances seem to take place on a yearly basis.

Diesel technology has been around for decades but has become increasingly popular in the past five years. Aftermarket companies are stepping up to the plate with performance products because Diesel owners are seeking more power. While making more power is great, knowing how your Power Stroke engine functions would also be a great benefit. While there are some Diesel shops around, only a few seem to understand the Power Stroke engine and how it functions.

This book is composed of information and illustrations that give you an idea of how this engine functions. While everything learned at my shop cannot be covered here, this book will give you some insight when problems do arise. It can also be a reference to owners who want to learn more. I hope this book is informative and as enjoyable to you as writing it was for me.

# ACKNOWLEDGMENTS

First and foremost, I want to thank God for giving me the ability to pursue this project. I also want to thank CarTech for giving me this opportunity. It has been a pleasure to work with Scott Parkhurst.

I would like to say thanks to my wife, Amber, for being so understanding and helpful in supporting me with this project. There were many hours spent away from each other but only because she believes in me.

Many people have influenced my life, such as my grandfather Grant Hienefield, who always encouraged me to never give up. Now that he has passed, I often look back and see how I took his words of encouragement for granted. I also want to thank my friend Greg Finnican who thought a Diesel book was a great thing to pursue and encouraged me to write it.

I would like to say that if you want to find in-depth technical information from Ford Motor Company, do not waste your time. The research in this book has been on my own from countless hours of owning a Diesel shop.

Thanks to all of the aftermarket manufacturers that supplied parts for the project. Diesel engines can be very expensive to rebuild and without their help, this book would not be possible. I have certainly enjoyed the opportunity to establish great relationships with people who devote themselves to making quality parts.

# What is a Workbench® Book?

This *Workbench®* Series book is the only book of its kind on the market. No other book offers the same combination of detailed hands-on information and revealing color photographs to illustrate engine rebuilding. Rest assured, you have purchased an indispensable companion that will expertly guide you, one step at a time, through each important stage of the rebuilding process. This book is packed with real world techniques and practical tips for expertly performing rebuild procedures, not vague instructions or unnecessary processes. At-home mechanics or enthusiast builders strive for professional results, and the instruction in our *Workbench®* Series books help you realize pro-caliber results. Hundreds of photos guide you through the entire process from start to finish, with informative captions containing comprehensive instructions for every step of the process.

The step-by-step photo procedures also contain many additional photos that show how to install high-performance components, modify stock components for special applications, or even call attention to assembly steps that are critical to proper operation or safety. These are labeled with unique icons. These symbols represent an idea, and photos marked with the icons contain important, specialized information.

Here are some of the icons found in *Workbench®* books:

***Important!–*** Calls special attention to a step or procedure, so that the procedure is correctly performed. This prevents damage to a vehicle, system, or component.

***Save Money–*** Illustrates a method or alternate method of performing a rebuild step that will save money but still give acceptable results.

***Torque Fasteners–*** Illustrates a fastener that must be properly tightened with a torque wrench at this point in the rebuild. The torque specs are usually provided in the step.

***Special Tool–*** Illustrates the use of a special tool that may be required or can make the job easier (caption with photo explains further).

***Performance Tip–*** Indicates a procedure or modification that can improve performance. Step most often applies to high-performance or racing engines.

***Critical Inspection–*** Indicates that a component must be inspected to ensure proper operation of the engine.

***Precision Measurement–*** Illustrates a precision measurement or adjustment that is required at this point in the rebuild.

***Professional Mechanic Tip–*** Illustrates a step in the rebuild that non-professionals may not know. It may illustrate a shortcut, or a trick to improve reliability, prevent component damage, etc.

***Documentation Required–*** Illustrates a point in the rebuild where the reader should write down a particular measurement, size, part number, etc. for later reference or photograph a part, area or system of the vehicle for future reference.

***Tech Tip–*** Tech Tips provide brief coverage of important subject matter that doesn't naturally fall into the text or step-by-step procedures of a chapter. Tech Tips contain valuable hints, important info, or outstanding products that professionals have discovered after years of work. These will add to your understanding of the process, and help you get the most power, economy, and reliability from your engine.

# Introduction

It has been more than 15 years since the name "Power Stroke" was introduced to the automotive world. I am confident to say that if you were to ask anyone on the street what a Power Stroke was, they would have to relate it to the Ford mid-size truck. This was great marketing by Ford and the Navistar Corporation when this Diesel engine was introduced. But, the Power Stroke quickly gained a reputation for being a truly loved workhorse. It offered great horsepower and torque along with fuel efficiency. This was an engine that quickly earned respect for its reliability. If you owned a Power Stroke, people knew you were serious about your pulling needs. Dodge and General Motors also had reputable Diesel engines, but nothing to compare with the pulling power of the Power Stroke. Often, owners of other brands were trading their vehicles for one of these Ford trucks.

This engine set the benchmark for Diesel technology that is still used today. When this powerplant was introduced, it was not your typical Diesel engine. It had computer-controlled engine management that had never been seen before. Turbocharging was a standard option, along with a direct-injection fuel system that, when combined, made for a very efficient and potent package. Regarding the internals of the engine, a lot of engineering firsts had taken place. The cylinder heads, pistons, and fuel system had all undergone major changes that are still being used today. There were so many things that Diesel engineers had discovered and introduced as a great package, it took the Ford truck a step above the competition.

In the past twenty years, having a Diesel engine in a mid-size truck has gained tremendous popularity. Whether it is in a 3/4- or 1-ton chassis, the Diesel option is becoming an industry standard. With this increasing popularity, many technological advances have brought the Diesel where it is today. I am sure we all remember the Diesel engine the way it used to be. The Diesel in the late 1980s and early 1990s was a rattling and clacking bucket of bolts by comparison to today's smoother running designs.

*This was a new look under the hood of the Ford truck. Now labeled "Power Stroke," it proved to be an amazing workhorse from the Navistar Corporation. With many technological advances this engine brought about a new era in Diesel engine technology.*

*In 1993, the 7.3 indirect injection (IDI) Ford truck was offered with a turbo. Other manufacturers were offering turbocharged Diesels, so to keep up with demand Navistar offered one on the last year of indirect injection models. This was offered for one year only, until the engine was replaced by the Power Stroke in 1994.*

*This badge was carried on the Ford truck with a turbo under the older 7.3 IDI engine.*

*This badge was a great marketing strategy for Ford and Navistar. Power Stroke, written in italics, was a name that anybody could remember what it stood for.*

The Diesels of the past were not really sought after among consumers. The only good thing they offered was longevity. The maintenance costs were somewhat high and the fuel economy poor. The horsepower and torque were not that great when compared to the same-size gasoline engines. When considering a Diesel truck, people were often afraid of the unknown. There were not too many dealerships that could confidently repair the vehicle if there was a problem.

This is where Ford took bold steps into the Diesel truck market. This engine was produced by International Harvester Corporation. It was 6.9 liters in displacement (420 ci). International actually started developing this engine in 1978, but it was not offered as an option in Ford F-series trucks until 1983. The 6.9 was noted for being simple in design and making great low-speed power. It produced 338 ft-lbs of torque at 1,400 rpm and 170 hp at 3,300 rpm. It was definitely not in the ballpark of today's Diesels, but it was known for its dependability.

In 1988, the 6.9 was replaced by the 7.3-liter version. The engines were very similar; the only changes being the cylinder bore diameter and compression ratio. By increasing the bore from 4.00 to 4.11 inches and compression from 20.7:1 to 21.5:1, the 7.3 was born. Now this engine made 185 hp at 3,000 rpm and 360 ft-lbs at 1,400 rpm. In 1993, International produced a turbocharged version, which made 190 hp at 3,000 rpm and 388 ft-lbs at 1,400 rpm. This was done to stay competitive with the new General Motors and Dodge engines that were also turbocharged. This engine was also known for its reliability, but short-lived by being replaced with the "Power Stroke."

The Power Stroke engine is a totally different animal from earlier Diesels. Ford offered the Power Stroke in two versions. The first was the 7.3-liter from 1994 until 2002. The second version was a 6.0-liter offered from 2003 until 2007. These engines are very unique in design and do not interchange with any other Diesel offered in a Ford truck. On these engines turbocharging was standard. These engines are outfitted with direct injection (DI), where fuel is injected directly into the combustion chamber.

When comparing the two versions of the Power Stroke, the 7.3 has the most respect among Ford owners. This engine carries the same bore and stroke of the original 7.3 DI with a totally different block design. The compression ratio was lowered to 17.5:1 and had a totally redesigned fuel system. The 7.3 DI makes 275 hp at 2,800 rpm and 525 ft-lbs of torque at 1,600 rpm.

With tightening emissions standards, Navistar created the 6.0 DI. This engine still functio ns like the 7.3 DI only in a smaller package. The bore size is 3.74 inches with a stroke of 4.13 inches. This engine makes 325 hp at 3,300 rpm and 570 ft-lbs of torque at 2,000 rpm.

The Power Stroke engines have proven to be reliable, durable powerplants when they are well maintained. They are responsive to enthusiast upgrades that are well-engineered and can deliver impressive fuel economy when driven conservatively. Ford's Power Stroke Diesels are not necessarily simple engines, but their performance justifies rebuild efforts and expenses. If they are rebuilt carefully, both the 7.3 and 6.0 can provide hundreds of thousands of miles of reliable service.

*As Power Stoke continued, so did government standards for Diesel engines. In order to meet those demands Navistar produced the 6.0 version of the Power Stroke.*

# SYSTEMS OVERVIEW

In 1994, Ford introduced the new breed of Navistar engine to its truck line called the Power Stroke. The Power Stroke was still a 7.3 engine that was redesigned and now known as a DIT (Direct Injected Turbo).

The engine came standard with a turbo made by Garrett. At first, the engine was turbocharged without an intercooler. If you look on the turbo outlet of the compressor side of the turbo you can see that a "Y" pipe is connected to direct air from the turbo into the intake manifold. The reason for no intercooler was the size of the turbo.

In 1999, Ford introduced an intercooler for the same 7.3 DIT engine. The intercooler was put into production on the Ford trucks because there was a change in the turbo. The compressor wheel was larger and the change to a bigger turbo brought more boost and higher exhaust gas temperatures. The new turbo was capable of maximum boost of 28 psi, which needed to be controlled.

Garrett installed a wastegate, mounted in the turbine side of the turbo. There is a wastegate actuator mounted on the compressor side of the turbo, connected to the wastegate by a rod. So when the compressor housing makes boost beyond what is needed, the actuator opens the wastegate.

In 2003, Ford introduced the 6.0 DIT engine for its Diesel trucks with a turbo known as variable geometric turbo (VGT) made by Garrett for this engine.

As engine speed rises, the powertrain control module (PCM) commands a solenoid to open at a desired rate to change the geometry of the "vanes" in the turbine housing. This is used to operate from no boost at idle to full boost at wide open throttle (WOT).

## Fuel System

Before the Power Stroke, Diesels were generally defined as indirect injection (IDI) engines. When the Power Stroke was introduced by the International Corporation new terminology came into use: direct injection (DI). Direct injection is used on the 7.3- and 6.0-liter engines. The use of computer control and direct injection made for less emissions and more power, bringing Ford's Diesel engines into the modern age.

### *Direct Injection*

With DI, Diesel fuel is pressurized electronically at the injector of the cylinder that is being fired. Other manufacturers were using electronic systems for Diesel injection, but they were also using an external pump to pressurize the Diesel fuel. The International Corporation used engine oil under high pressure along with various sensors and lines to make Diesel fuel pressure at the injectors.

### *7.3 Fuel Supply*

Fuel is stored in the tank and must be pressure-fed to the fuel injectors. On the 7.3 from 1994 to 1997, fuel came from the tank through fuel lines to a mechanical engine-driven pump, which is located in the central valley of the engine behind the turbocharger.

One side of the pump is the inlet, used to suction fuel from the tank and deliver it to the fuel filter basket. The fuel pressure coming into the fuel filter basket is anywhere from 3 to 7 psi. On the other side, at the fuel filter basket is a water separator drain.

When entering the second stage, fuel is compressed further and sent to the cylinder-head–mounted fuel injectors at 40 psi.

#### *6.0 Fuel Supply*

The 6.0 fuel system is very similar to that on the later 7.3. This system has an electric pump known as a horizontal fuel conditioning module (HFCM). The HFCM is frame-mounted, and is controlled by the engine's computer. The condensation drain (water separator) is integrated into the pump, as is a serviceable pre-filter.

## Oiling System

The Power Stroke oiling system is not radically different than that of other internal combustion engines. The only major design differences are the high-pressure path required by the high-pressure oil pump (HPOP) and the oil cooler.

There are differences in the HPOP systems of the 7.3 and 6.0. Even though their functions are similar, the differences are worth noting. There are two system types: the low-pressure system and the high-pressure system.

#### *7.3 Low-Pressure System*

The oil pump in the 7.3 engine is of a "gerotor" design. As the crankshaft spins, the gear on the front turns the rotor (vane) inside the oil pump. Oil is drawn through the pick-up screen mounted in the oil pan.

When oil reaches the pump it is then pressurized and sent through the front cover. The front cover has two passages by which the oil enters the block. One passage leads to the reservoir for the HPOP. The other passage sends oil to the oil cooler, which is mounted on the side of the engine.

#### *7.3 High-Pressure System*

In order for the oil to enter the HPOP reservoir, it must pass through a check-ball assembly integrated into the front of the block. During cold starts, the HPOP receives unfiltered oil through the check-ball passage until reaching a nominal operating temperature.

When oil enters the HPOP, it is pressurized and sent through hoses that run to both cylinder heads.

Oil then travels through the high-pressure passages to surround each injector. When the injector is fired, the oil drives the internal piston down inside the injector body. After the piston in the injector is driven down and the piston returns to the top the oil is "spit" out of the injector through an orifice mounted on the side of the top of the injector.

The oil then drains back through holes leading to the oil pan.

#### *6.0 High-Pressure System*

The 6.0 HPOP is housed in the rear of the motor. The crankshaft, camshaft, and HPOP are gear driven from the rear of the engine. The front of the engine has no gears other than the crankshaft, which turns the engine oil supply gerotor pump.

As the HPOP spins from the rear geartrain, it pressurizes oil received from the reservoir. The pressurized oil exits the HPOP into a pump discharge line.

The design of HPOP system in the 6.0 DI is more efficient than that in the 7.3 DI. If any leaks occur in the system they are contained by the crankcase.

If a leak occurres on a 7.3, it can be detrimental. If the high-pressure hoses come apart oil can spray everywhere, cause engine damage, and make a huge mess. The design of the 6.0 system came about by problems faced with the 7.3. The HPOP system on the 6.0 is more of a task to service, but most of the time it's trouble-free.

## Cooling System

The cooling system of the Power Stroke engine does employ one component you may not be familiar with: the degas bottle. This small tank is mounted separately from the

***This is the HPOP on the 6.0 Power Stroke. It is located at the rear of the engine underneath the turbo. The injection pressure regulator is located at the top right of the cover.***

radiator and serves as the service point for the cooling system. The service cap is mounted on the degas bottle, as it is mounted higher than the radiator (which has no cap).

In general, the cooling system of the 7.3 engine is reliable. Even in the event of adding aftermarket products (such as programmers and upgraded turbos), the coolant system has proven itself to be efficient. Typical wear parts (such as water pumps and thermostats) are to be expected, but it's rare to experience a head gasket repair or oil cooler failure.

#### *6.0 Engine Cooling System*

The 6.0 engine utilizes many of the same components as the 7.3 engine.

One thing that you need to address is the exhaust gas recirculation (EGR) cooler. In 2004, EGR valves became mandatory. The EGR systems did pose a problem when this engine was introduced. From 2004 to 2006, Navistar had some serious complaints with the EGR systems on the 6.0 engines. Additionally, head gaskets became a problem and plagued the 6.0 engine. Blown head gaskets were overheating engines and destroying them.

One head gasket problem was the material, but there are only four bolts per cylinder to secure them (compared to the six bolts per cylinder of the 7.3). Hard use over extended periods placed serious pressure on the head gaskets.

There are two solutions for the problematic EGR cooler. One is to totally eliminate the EGR from the engine. Keep in mind that this is for off-road use only. If you remove this for highway use, it is not emissions compliant.

The second solution is to replace the EGR cooler. Bulletproof Diesel sells a stainless-steel EGR cooler with a more rigid design, which offers more strength and cooling efficiency. Using this product is a great option instead of having to replace the factory cooler and carries a lifetime warranty.

### Cylinder Heads

In the 7.3 cylinder head, you have one intake valve and one exhaust valve. They both are the same size, 1.68 inch diameter. From 1994 until 2002, the 7.3 cylinder head remained the same. The injectors and glow plugs are under the valve cover.

When the 6.0 was released in 2003, many changes had taken place, including the cylinder heads. This was the first head from Navistar with four valves per cylinder (two intake valves and two exhaust valves). Valves measure at 1.33 and the exhaust valves measure 1.10 inches in diameter.

The cylinder head took on a totally new configuration to help meet

***Notice the difference between the 6.0 cylinder head (left) and the 7.3 cylinder head (right). There are four valves per cylinder versus two valves, but size does matter. The 6.0 DIT has 79 fewer cubic inches. That is almost a cylinder and a half less than the 7.3 DIT. To keep up with emissions demands, smaller packages were introduced.***

***The rocker arms on the 6.0 DIT are attached to the rocker carrier. This is a complete assembly with all rocker arms that attach it to the cylinder head. It is held in place with Torx-head bolts attaching to the cylinder head.***

the emissions standards. To increase port swirl, the valves were positioned in a twisted configuration in relation to the cylinder. The Diesel engine is dependent upon port swirl and with this valve layout, the combustion process is more refined and efficient.

Glow plugs from the 6.0 now have removable sleeves. On the 7.3 engine, the glow plugs thread into the top of the cylinder head in a "cast" boss leading to the combustion chamber.

As for the 6.0 engine, the glow plugs are sealed by a stainless-steel sleeve. The sleeve seals off the injector and the combustion chamber from coolant.

Because the 6.0 cylinder head has four valves per cylinder, two intake valves have to be opened at the same time along with two exhaust valves. The rocker arms are different from the 7.3 in relation to design and length. Also, the rocker arms incorporate a bridge linking the two.

Something else to note is the injector hold-downs. On the 7.3 engine the injector has a hold-down that is on the injector body. There are two bolts positioned from top to bottom where the injector sits in the cylinder head. The top bolt is installed into the cylinder head and tightened to its specific torque specification. Then the injector is installed into the cylinder head and the hold-down is positioned over the bolt already installed. Once it is in position the other bolt can be installed.

As for the 6.0, the injector hold-down is much easier to deal with. The hold-down is shaped like a horseshoe and is removable from the injector body. One caution: take care when installing the injector in the 7.3 or 6.0. Make sure to grease the O-rings on the injector along with the injectors cup in the cylinder head. Then install the injector and press it into the head by hand or with a soft rubber hammer. Do not strike the injector with a blunt object because this can damage the injector.

## Exhaust System

The "up pipe" carries the exhaust gas from the manifold to the turbo. This pipe connects to the turbo through a "Y" pipe. The Y is cast iron and directs exhaust flow to the turbine chamber in the turbo.

For the 7.3 engine, before the exhaust enters the "down pipe" it must pass through an exhaust backpressure valve (EBV), which is a flap in the end of the turbine housing on the exhaust exit side of the turbo. In simple terms, the EBV flap can open or close to restrict the exhaust flow.

The EBV is used by the PCM of the engine to help aid in warm-up in cooler temperatures. This provides more heat to the coolant for interior heat when the outside temperature is below 45 degrees F and the oil temperature is below 167 degrees F.

### *6.0 Engine Exhaust*

In order to meet the demands of tighter emissions in 2003 Navistar introduced the 6.0 engine, the government demanded that manufacturers reduce the amount of nitrogen oxide in Diesel engines.

The 6.0 engine incorporates an exhaust gas recirculation (EGR) valve.

Selective Catalyst Reduction (SCR) is something new and has just been released for Ford trucks. The SCR system uses urea (odorless, colorless, non-toxic substance found in urine of mammals) as the catalyst.

The urea used in the exhaust system is in liquid form. This liquid can be purchased at most local parts stores as Diesel exhaust fluid (DEF). DEF is made from a concentration of liquid urea plus a percentage of deionized water.

*The injector hold-down of the 6.0 DIT is in the shape of a horseshoe and fastens to a machined pad in the cylinder head. The machined pad of the hold-down is on the edge of the valve cover gasket rail in the cylinder head toward the exhaust side.*

A truck has two tanks inside the fuel filler door. One is for the Diesel exhaust fluid (DEF) (blue lid) and the other is for Diesel fuel (green lid). When using urea, there is no need to use an EGR valve.

But the draw back is that now the engine makes nitrous oxide. So to eliminate that, urea is injected through a nozzle downstream in the exhaust system somewhere shortly after it exits the turbo. The urea reacts with the nitrous oxide, turning it into ammonia. The ammonia enters the catalytic converter, where it is separated into nitrogen and water.

The new SCR system is the best choice for reduction in emissions for Diesel engines. Yes, there is an expense of the DEF, but in return you make more power and obtain better fuel economy.

## Electrical System

In 1994, when the Power Stroke engine was introduced, it was the first computer-controlled Diesel for the mid-size truck. The engine design was known as a HEUI (hydraulically actuated, electronic controlled unit injector). There were some other Diesel engines on the road that had some functions that were computer controlled, but not the entire engine.

## Glossary of Terms

### *Accelerator Pedal Position (APS) Sensor*

The APS is attached to the accelerator pedal and is a part of the pedal assembly. The engine has no throttle cable attached to the accelerator pedal. As the pedal is depressed to the floor, the sensor sends a voltage reading to the PCM. The PCM reads the voltage and determines that if the accelerator pedal has been depressed then more fuel needs to be given to the engine. This type of system is also known as "fly-by-wire."

### *Barometric (BARO) Pressure Sensor*

This sensor is located behind the instrument panel. The engine's computer references this sensor to indicate changes in altitude, which affect the quantity of fuel and changes in the timing.

### *Camshaft Position (CMP) Sensor*

The CMP is also called a "hall effect" switch. It uses magnets to signal position. The camshaft has a "tone" wheel. This wheel is made up of slots and teeth. As the teeth of the tone wheel pass the magnets of the sensor, voltage is produced. When the slots pass by the sensor, magnetism drops off and crankshaft position can be established and referred to by the computer.

The PCM uses the frequency from the crankshaft position (CKP) sensor and CMP to determine engine speed and position. The PCM also conditions the signal from the CMP to be used as the tachometer reference in the instrument cluster.

### *Crankshaft Position (CKP) Sensor*

This sensor receives pulses from a tone wheel mounted on the crankshaft of the engine. It works very similarly to the function of the CMP. The CKP determines crankshaft speed, position, and acceleration. This information is used as input to the computer for misfire detection and the control of fuel to the cylinders through the injector driver module (IDM).

The CKP was not used for the 7.3 engines. On the crankshaft of the engine is a slotted target wheel, 58 evenly spaced teeth, and a slot on the wheel that is the width of two teeth that have been removed (known as the SYNC gap). As the crankshaft spins, the teeth pass the magnetic field of the sensor which causes a voltage frequency. The voltage frequency continues until it passes the minus-2 slot that causes a drop in frequency. The frequency changes and the minus-2 slot is used by the PCM to determine engine speed and position relative to top dead center (TDC).

*This plate with windows machined into it goes over the front of the camshaft drive gear. The windows create the "hall effect" voltage that the CMP uses to tell the computer to pulse the injectors. If you look closely you can see that one window is larger than the other ones. This is used to reference the computer for the number-1 cylinder as it rotates.*

### *Engine Coolant Temperature (ECT) Sensor*

The ECT sensor is used on both the 7.3 and 6.0 engines. The sensor tells the computer the temperature of the engine coolant. The computer then uses this information to control functions of the engine. The ECT sensor is located in the thermostat housing.

### *Engine Oil Pressure (EOP) Sensor*

This sensor is used for dash instrumentation only. The EOP is not read by the PCM. The EOP switch closes when oil pressure reaches 5 to 7 psi. The normal engine oil pressure in the 6.0 engine ranges from 5 to 70 psi.

### *Engine Oil Temperature (EOT)*

The EOT works much like a coolant temperature sensor. PCM reads the resistance in order to determine oil temperature.

### *Exhaust Back Pressure (EBP)*

The EBP is used to measure the exhaust pressure inside the exhaust manifold. The readings taken by the PCM from the EBP sensor will be to control the EBP valve and to help aid in warm up. Second, the EBP voltage signal indicates to the PCM the performance of the turbo.

### *Exhaust Gas Recirculation (EGR) Valve*

When certain parameters of the PCM are reached such as engine speed, coolant temp, oil temp, etc., the EGR is commanded to open. The purpose of this actuator is to open and allow exhaust gas to enter the intake manifold. How far it opens is also determined by the PCM from parameters of the engine at that time.

This actuator was introduced on the 6.0 engine in order to reduce emissions.

### *Fuel Injector Control Module (FICM)*

In order for the injectors to pulse from the signal sent to the computer from the CMP sensor, the computer must use a device to pulse the injector, the FICM on a 6.0 (or the injector driver module, IDM, on a 7.3). In simple terms, this is a transformer box. For the injectors to pulse under great loads takes more than 12 volts. The voltage needed to pulse such an injector is around 120 volts.

### *Glow Plugs*

Power Stroke engines use a glow plug system. Like spark plugs in gasoline engines, these plugs ignite incoming fuel to start the engine. Unlike spark plugs, glow plugs maintain heat and ignite the fuel when compression levels reach a combustible point. The hotter the glow plugs are, the lower this compression point needs to be.

A "Wait to Start" light in the dash alerts the driver to hold off turning the key until the glow plugs have reached their predetermined heat range.

With growing technological advances, 6.0 engines use a glow plug control module (GPCM). This module works like the glow plug relay on 7.3 engines. The biggest difference is that it can control the glow plugs individually.

The GPCM module is controlled by the ECM, and if a glow plug failure occurs, it alerts the computer of its functionality for diagnostics.

On both engine systems, the computer may even cycle the glow plugs when the engine is running. It all depends on the temperature readings given to the computer. After the engine has warmed to normal operating temperature and has been turned off, the glow plugs may not be used at all to restart. It depends on the feedback the ECM receives from the sensors.

### *Injection Control Pressure (ICP) Sensor*

The ICP tells the computer what pressure is being produced by the HPOP. Then the computer controls the regulator on the demand for the desired oil pressure. It is much like the one used for the 7.3 engines. The ICP is a variable capacitance sensor that is the "eyes" for the PCM to determine pressure being produced by the HPOP.

On the 7.3 engine, there is an injector driver module (IDM). Instead of using an IDM on the 6.0 engine, International used the fuel injector control Module (FICM).

### *Injection Pressure Regulator (IPR)*

The IPR is mounted on the HPOP and controls the amount of oil pressure being generated in the HPOP. It controls how much oil being delivered by the engine oil pump enters the HPOP to be pressurized. Basically, the IPR is an adjustable regulator that is controlled by the PCM to help maintain proper needed pressure of the HPOP.

### *Injectors*

The injector of the Power Stroke is where the "HEUI" name is derived. The injector is actuated by high-pressure oil from the HPOP. The injector has a solenoid mounted on top and when energized by the injector drive module (IDM) uses the high pressure oil to squeeze Diesel fuel through the injector's tip. The solenoid takes 100 volts and 7 amps to energize.

The 6.0 injectors are much like the ones used for the 7.3 engine. They both have to use high-pressure oil from the HPOP in order for the plunger of the injector to be driven down. In the 7.3 engine, high-pressure oil is delivered to the injector by a "barrel" cast into the cylinder head. The high-pressure oil surrounds each of the injectors in the cylinder head. When the injectors coil is energized, high-pressure oil is allowed in to drive the plunger down.

The 6.0 injector is a lot smaller than the 7.3 with fewer O-rings.

### *Intake Air Temperature (IAT) Sensor*

It works very similar to the engine oil temperature (EOT). Inside is a thermistor that produces resistance to the PCM to help determine temperature. As the intake air temperature increases, the resistance decreases. The IAT is mounted in the intake air cleaner to provide the PCM with information on the outside air temperature. The PCM uses this information to determine the need for the use of the exhaust back pressure (EBP) control to aid in warming the engine.

### *Mass Air Flow (MAF) Sensor*

This sensor is used to determine the amount of air that enters the engine. The MAF is mounted between the air filter and turbo inlet. Inside the MAF is an element that is heated. As outside air is pulled in by the engine, it passes across the heated element inside the MAF. As the air passes across the heated element, it cools the element. This generates a signal back to the PCM to determine fueling requirements in proportion to how much air is being taken in by the engine.

### *Manifold Absolute Pressure (MAP) Sensor*

This sensor is also used in gasoline engines to measure atmospheric pressure inside the intake manifold. The difference in this application is the use of a turbocharger. As the accelerator pedal is pushed down and the pressure rises inside the intake manifold, the computer makes changes in the fuel distribution to the cylinders.

### *Power Control Module (PCM)*

The PCM contains three microprocessors: the engine, chassis, and transmission. The three microprocessors work together from information provided by various sensors from the engine, transmission, and chassis to determine a fueling strategy for the engine. The PCM sends a 5-volt reference to the sensors, except for the crank and cam sensors, which generate their own voltage for the PCM. The PCM uses nine sensors from the engine microprocessor and four from the chassis microprocessor to fine-tune the engine.

***The 7.3 DI injector is much bigger than the 6.0 DI injector. The location of the O-rings is for the separation of the chambers in the injectors housing. The high-pressure oil enters the top and Diesel fuel enters the bottom.***

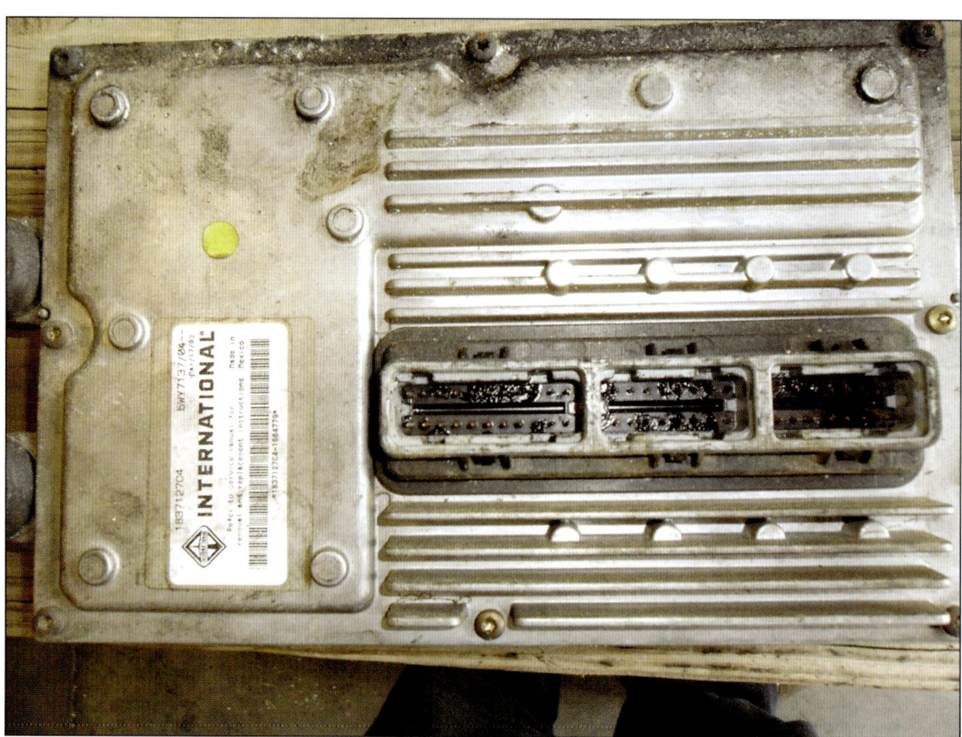


***In 2003 on the 6.0 Power Stroke, the IDM became known as the FICM. It is located behind the coolant tank at the rear of the driver-side valve cover.***

CHAPTER 2

# 7.3-Liter Removal and Disassembly

Because Diesels are a different breed of internal combustion engine, the common practice was to purchase a used engine from the local salvage yard instead of rebuilding one. But this has become a much harder task. Salvage engines were an easy replacement but they have become scarce. Not only are they getting harder to find but also the condition and mileage of the used engine may become a factor. Over the years of production, aftermarket companies have begun manufacturing parts necessary for their rebuilds.

To rebuild your Power Stroke, you will need some basic hand tools such as sockets and wrenches and also some specialty tools. Most of the hand tools are metric— 8, 10, 13, 15, 16, and 18 mm—and include sockets for 1/4-, 3/8-, and 1/2-inch-drive ratchets. Tool suppliers like Mac Tools, Snap-on, and Matco may have the specialty tools required to properly rebuild a Power Stroke.

For the 7.3 engine removal and disassembly, an older body style (1994 to 1997) was chosen for the step-by-step coverage in this book. Plenty of these trucks are still on the road today and many of them are in need of a rebuild or are certainly approaching that point. The only major issue in the engine removal of these year models is the radiator support. From 1999 to present, the radiator supports can be removed for easier access. But in the earlier 1994 to 1997 models, you simply have to work around the radiator support. The following step-by-step photos give you an idea of what is involved.

## Begin Removal

### 1 Disconnect Battery

*The first and most important thing to do is to disconnect the battery. The passenger side is exposed so start with this battery first. Remember to isolate the positive battery cable away from other metal objects until you can disconnect the driver side.*

## 2 Remove Intake Scoop

*In order to disconnect the driver-side battery you have to remove the air cleaner intake scoop. This is usually held in place by a bolt with an 8-mm socket head.*

## 3 Drain Oil

*Use a 3/4-inch wrench and drain the oil. If you perform your own oil changes this shouldn't be a problem.*

*Professional Mechanic Tip*

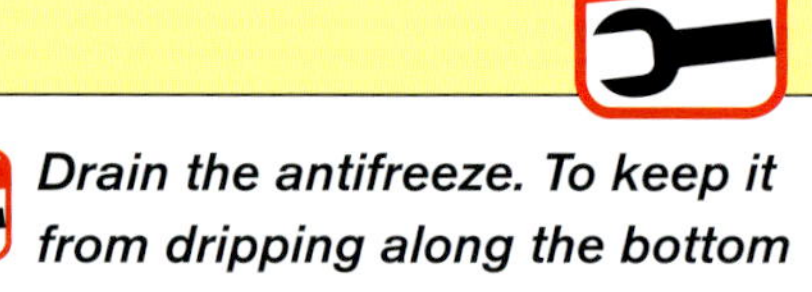

## 4 Drain Antifreeze

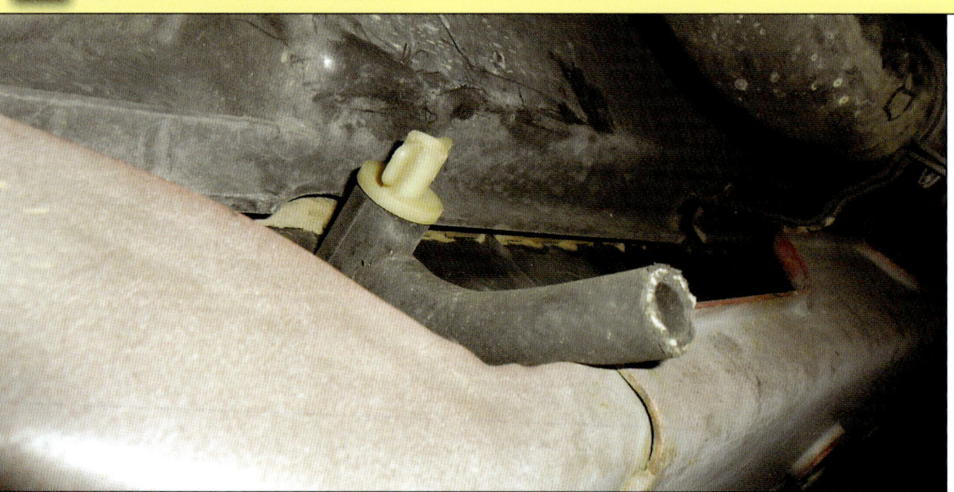

*Drain the antifreeze. To keep it from dripping along the bottom of the lower radiator support, I install a small piece of 3/8-inch rubber fuel line to the drain nipple. This makes it easier and less messy to aim for the drain pan.*

## Air Inlet System

### 1 Remove Air Cleaner Lid Bolts

*Use a 1/4-inch-drive rachet with an extension to remove the plastic bolts that hold the top of the air cleaner lid in place. The 1/4-inch extension fits into the top of the plastic bolts.*

### 2 Loosen Hose Clamp

*Loosen the clamp on the rubber hose coming from the turbo inlet pipe to the air cleaner assembly.*

### 3 Remove Bolts

*Use a 10-mm socket to remove the bolts that hold the lower air cleaner box to the inner fender.*

## 4 Disconnect IAT Sensor

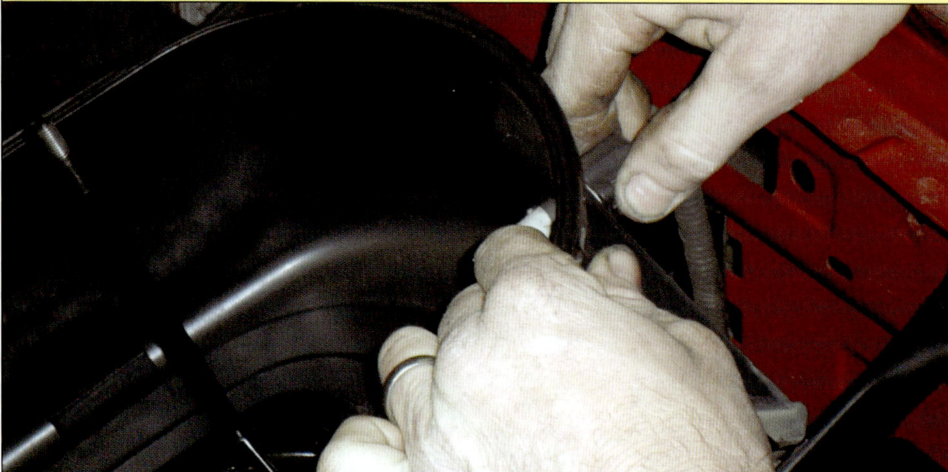

*Disconnect the intake air temperature (IAT) sensor and remove the lower air cleaner assembly.*

## 5 Remove Battery Hold-Downs

*Use an 8-mm socket to remove the battery hold-down bolts. Remove both batteries.*

## 6 Remove Radiator Hose

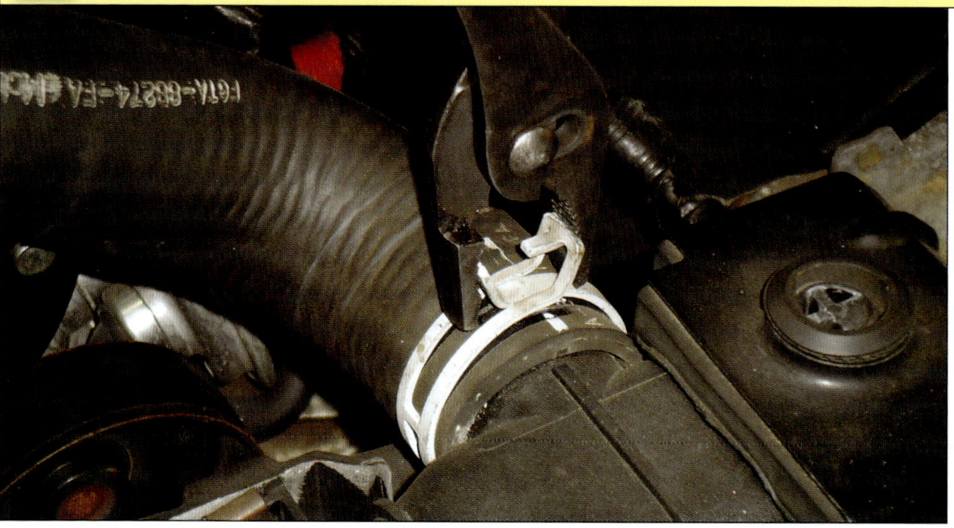

*Remove the upper radiator hose by squeezing the top of the clamp together with a pair of pliers.*

## Front Belt Drive Accessories and Radiator

### 1 Loosen Clutch Fan

*While the serpentine belt is on the motor, take this opportunity to loosen the clutch fan from the water pump. The belt serves as an extra hand in the loosening process. (Remember, this should be done with a special wrench you can obtain from the local parts store for this application.) At this time do not fully remove the clutch fan from the water pump.*

Professional Mechanic Tip

PRO TIP

### 2 Turn Belt Tensioner Pulley

PRO TIP *After loosening the clutch fan from the water pump, you can remove the drive belt by turning the belt tensioner pulley. The tensioner pulley is on a spring-loaded cam that applies tension to the belt. Place a 1/2-inch-drive rachet with a 15-mm socket on the bolt in the center of the pulley and turn.*

### 3 Loosen Belt

*This is a view from the top of the radiator with the rachet and socket in position to loosen the belt.*

## 4 Remove Coolant Hoses

*Use a pair of pliers to remove the clamps on the hoses at the coolant degas bottle. Remove the coolant hoses from the degas bottle.*

## 5 Remove Degas Bottle

*Remove the bolts that hold the degas bottle to the passenger-side inner fender wheel using a 10-mm socket.*

## 6 Remove Clutch Fan and Fan Shroud

*With an 8-mm socket, remove the two bolts at the top of the radiator that hold the fan shroud to the radiator. After removing these bolts, finish loosening the clutch fan from the water pump. Then remove the clutch fan and fan shroud together at one time.*

## 7 Remove Lower Radiator Hose

*Remove the lower radiator hose from the water pump housing.*

## 8 Remove Radiator

*Now the radiator can be removed. Using a 10-mm socket, remove the four bolts (two on each side) that hold the radiator to the radiator support.*

## 9 Disconnect and Remove Alternator

*With the radiator removed, more engine components are exposed. Disconnect the wiring from the alternator. Using a 13-mm socket, remove the alternator from the engine bracket.*

Special Tool

## 10 Remove Pulley Bolt

*Remove the idler pulley below the alternator on the engine bracket. This is done by removing the bolt in the center of the pulley using a size-50 Torx bit socket.*

## 11 Remove Pulley

*Next, the pulley from the tensioner can be removed. Using a 15-mm socket, turn the bolt in the pulley as though you were tightening it. The bolt in the pulley has a left-handed thread, so you are actually loosening the bolt.*

## 12 Remove Tensioner

*Using the same size-50 Torx bit used to remove the idler pulley, you can now remove the tensioner from the engine bracket.*

## 13 Remove Engine Bracket

*With the accessories and pulleys removed from the passenger-side engine bracket, you can now remove the engine bracket by removing the bolts with a 13-mm socket.*

## 14 Separate Connector

*The engine wiring harness connector is on top of the driver-side valve cover. The connector is held together with a bolt in the center. As you loosen the bolt in the center with an 8-mm socket, the connector separates but the bolt does not fall out; the bolt is encased in the connector.*

## 15 Remove Vacuum Pump

*The vacuum pump is on top of the driver-side engine bracket. Disconnect the vacuum line and remove the pump by removing the bolts that are located behind the pulley.*

## 16 Remove Pulley

*Use a power steering pulley remover and installer tool to remove the pulley so you can gain access to the bolts that hold the power steering pump to the engine bracket.*

## 17 Remove Pump

*Once the pulley is removed from the power steering pump, use a 13-mm socket to remove the bolts from the front of the pump. At this time you can lay the pump to the side without disconnecting any lines.*

## 18 Remove Bracket Bolts

*Use a 13-mm socket to remove the bolts that hold the driver-side engine bracket to the engine.*

## 19 Remove Down Pipe Clamp

*The exhaust down pipe that attaches to the turbo is right next to the firewall. At this time remove the clamp that holds the down pipe to the turbo and let it lay there.*

## 20 Loosen Down Pipe

*This is how close the down pipe is to the firewall. At this time the pipe does not need to be removed, just loosened.*

# Final Accessories

## 1 Remove Ground Cable

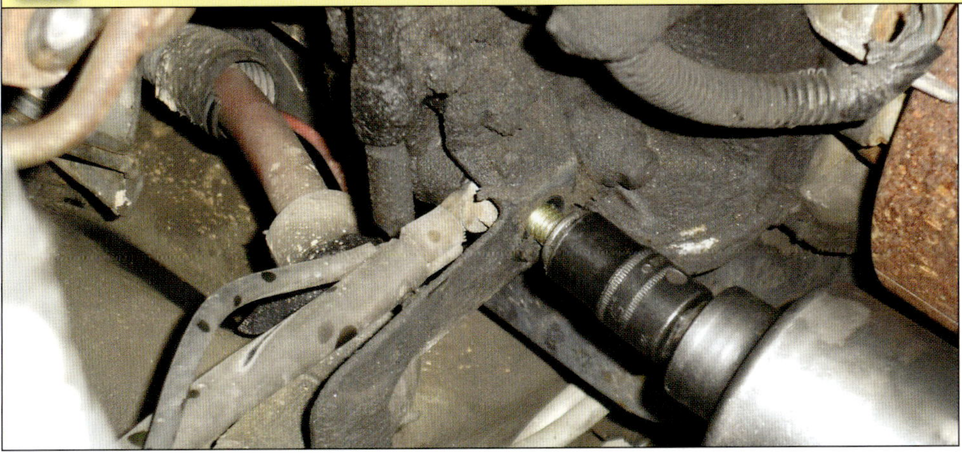

*At the front of the engine on the passenger's side near the bottom of the block is a stud that supports the positive battery cable bracket and the ground cable. Using a 15-mm socket, remove the nut from the stud for the bracket. Once the bracket is off you can pull the stud and remove the ground cable.*

## 2 Remove Inlet Tube

The aluminum air inlet tube of the turbo is attached to a bracket that mounts to the driver-side valve cover. Use a 10-mm socket to remove the inlet tube from the bracket.

*Professional Mechanic Tip*

PRO TIP

## 3 Remove Rubber Tube

PRO TIP Loosen the clamp of the rubber inlet tube at the turbo and remove the rubber tube. Here, where the rubber tube attaches at the turbo it has started to deteriorate and has started to come apart. After years of service, oil residue from the turbo tends to form a puddle in this bend of the inlet tube causing delamination.

## 4 Remove Motor Mount Studs

Looking underneath (inside) the crossmember, use a 19-mm socket to remove the nuts from the studs of the motor mounts.

## 5 Remove Starter

*Remove the wires from the starter and, using a 17-mm socket, remove the starter.*

## 6 Remove Bellhousing Bolts

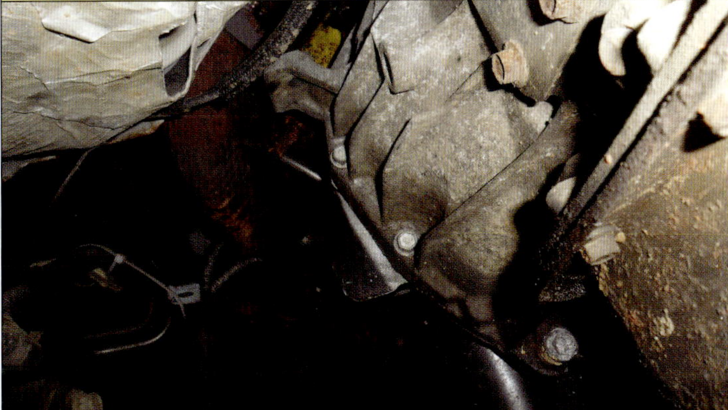

*Using a 13-mm socket, remove the bolts from the transmission bellhousing. Remove torque converter bolts (automatic transmission). Use a floor jack to support the transmission while the engine is being removed.*

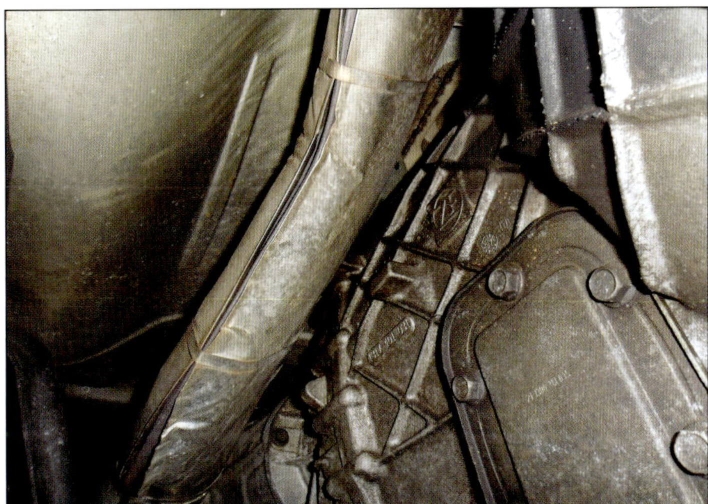

*The exhaust pipe from the turbo tightly fits between the transmission and the body. Notice that the factory squeezed the pipe to gain clearance.*

## 7 Remove Inlet Tube

*Loosen the clamps that hold the inlet tube from the turbo compressor housing to the intake of the engine. Remove the tube and at this time attach a chain to the engine lift brackets.*

### 8 Remove Pedestal Bolts

Remove the two bolts in the front of the turbo-mounting pedestal with a 10-mm socket.

### 9 Disconnect EBV

*Disconnect the exhaust backpressure valve (EBV) electrical connection on the driver's side at the bottom of the turbo compressor housing.*

## Turbocharger

*Professional Mechanic Tip* PRO TIP

### 1 Remove Front Turbo Housing Bolts

PRO TIP *Remove the bolts that connect the turbine housing of the turbo to the exhaust up-pipe manifold. This can be tricky! In order to make things easier, it may be in your best interest to remove the transmission because the bolts are nearly impossible to get out without breaking.*

### 2 Remove "Y" Manifold

*Here's a big challenge: The "Y" manifold has to come loose from the turbo before the turbo can be removed. The turbo has to come out before the engine can clear the windshield cowl panel.*

### 3 Remove Rear Turbo Housing Bolts

*After the "Y" manifold has been removed from the turbo, remove the bolts from the rear of the turbo-mounting pedestal. Then the turbo can be removed from the engine.*

## Engine Removal

### 1 Remove Engine

With a suitable hoist, remove the engine from the engine bay. Also, use a piece of 2 x 4 to wedge between the frame rails to support the transmission after the engine is removed.

*For 1994 to 1997 models, removal and installation of the engine is a tight squeeze due to the design of the engine compartment. This doesn't pose a real problem; just a little aggravation.*

### 2 Place Engine on Stand

*These engines weigh approximately 1,000 pounds. In order to support that kind of weight, you need a suitable engine stand rated for the appropriate capacity. This kind of stand can be very expensive because you have to be able to rotate the engine. An option is to build a support out of wood similar to this one so you can cradle the engine and remove many of the essentials, then move it to a budget engine stand to complete the teardown.*

## Begin Disassembly

### 1 Remove Lower Pipe Plugs

*Once the engine is removed and still on the engine hoist, remove the pipe plugs in the lower portion of the block to drain the antifreeze. Do this before placing the engine on a cradle or appropriate engine stand. It will make the teardown less messy especially when removing the cylinder heads.*

### 2 Remove Oil Cooler Pipe Plug

*In order to drain the antifreeze from the driver's side of the block, you need to remove the pipe plug at the rear of the oil cooler using a 5/16-inch square-headed socket.*

## 3 Remove Wiring Harness

*Remove the wiring harness from the engine.*

## 4 Disconnect Sensors

*Remove the bolt from the harness clamp at the lower portion of the water pump on the passenger's side. Disconnect the EBV sensor connectors and temperature sender connectors.*

## 5 Remove Relay Bracket Bolts

*Remove the bolts that hold the glow plug relay bracket to the passenger-side cylinder head.*

Professional Mechanic Tip

## 6 Remove Engine Harness

*Unplug the rest of the sensors and remove the engine harness. With the engine harness removed, the high-pressure hoses and fuel lines are now exposed and easier to access.*

## 7 Remove Oil Hoses From Cylinder Head

*Remove the high-pressure oil hoses from the fittings of the cylinder heads.*

### 8 Remove Hose Clamp

On the driver-side cylinder head, remove the high-pressure hose clamp at the cylinder head underneath the fuel filter restriction sensor.

### 9 Remove Oil Hoses From HPOP

Remove the high-pressure oil hoses from the HPOP.

## Fuel System

### 1 Remove Fuel Hoses

At the front of each of the cylinder heads are the fuel return hoses. Remove them from the fittings in the cylinder head.

Professional Mechanic Tip

## 2 Drain Fuel from Filter Basket

*Below the passenger-side cylinder head near the oil pan is the metal line from the fuel filter basket for the water separation drain. Place a drain pan under the metal pipe and open the water separator valve on the fuel filter basket. This allows the fuel to drain from the filter basket.*

## 3 Remove Fuel Inlet Tubes

*At the back of each of the cylinder heads are the fuel inlet tubes that branch together and join at a banjo fitting in the back of the fuel pump. Remove the fuel inlet tubes from the rear of each of the cylinder heads.*

## 4 Remove Fuel Inlet Branch

*Now you can remove the fuel inlet branch from the back of the fuel pump by removing the banjo fitting.*

## 5 Remove Fuel Supply Hoses

*Loosen the clamps of the rubber fuel supply hose at the fuel pump and rubber fuel return hose at the regulator attached to the fuel filter basket and remove the hoses.*

## 6 Remove Rubber Hose

*Remove the rubber hose at the bottom of the fuel filter basket that connects the water separation drain to the metal tube.*

## 7 Remove Fuel Filter Basket Bolts

*Remove the two bolts (one on each side) that hold the fuel filter basket to the engine block.*

*Important!*

## 8 Remove Fuel Pump and Fuel Basket

*Using a pry bar, gently lift up on the bottom of the fuel pump. Be careful, the fuel pump actuator rod that runs on the camshaft is sealed by an O-ring that can be stubborn to break free. If you get in too big of a hurry, the bottom of the fuel pump breaks off. Once the fuel pump is loose from the block, remove the fuel pump and fuel basket together at the same time.*

### 9 Remove Fuel Lines

*Now you can remove the fuel supply and return metal lines that are fastened to the cylinder head with a clamp.*

### 10 Remove Crankcase Breather

*Take a Phillips screwdriver and remove the screws that attach the engine crankcase breather to the valve cover.*

## Fuel Injectors

### 1 Remove Valve Cover Bolts

*On the driver side valve cover, remove the two nuts that hold the wiring harness support bracket to the valve cover. Then remove the rest of the bolts that hold the valve cover to the cylinder head.*

## 2 Remove Valve Cover Gasket

*The valve cover gasket of the 7.3 has the wiring harness for the injectors and glow plugs integrated into the gasket. In order to remove the gasket you need to disconnect the injector and glow plug harness.*

## 3 Remove Fuel Plug From Cylinder Head

*Before the injectors are removed, there are several preventative things to do to create less mess. At this time the cylinder head still has oil in the high-pressure gallies and fuel surrounding each injector. Start by removing the fuel plug in the rear of the cylinder head with a 1/4-inch ratchet to drain the fuel from the cylinder head.*

*Professional Mechanic Tip* PRO TIP

## 4 Drain Oil Reservoir

PRO TIP *The easiest way to drain the high-pressure oil reservoir from each cylinder head is to remove the plugs in the reservoir from the front and rear of the cylinder head. You can accomplish this with a 1/2-inch-drive ratchet, but be warned. Most of these plugs cannot be removed unless you apply some heat to the cylinder head first. Before removing the plug have a drain bucket handy, because there will be quite a bit of oil that is going to come out from the reservoir of each cylinder head.*

## 5 Remove Injector Bolts

*Using an 8-mm socket, remove the bolt in the center of the rocker arms below the injector. This bolt holds the injector to the cylinder head.*

## 6 Push Up on Injector Retaining Ring

*Once the bolt is removed, push up on the injector retaining ring. This frees the retaining ring from the other bolt in the retaining ring in the upper cylinder head. The retaining ring is made to slide on this bolt once the bottom bolt is removed.*

## 7 Remove Injector

*Now take a small pry bar under the retaining ring of the injector and pry up to remove the injector. Something I do is to mark which cylinder the injector came from.*

## 8 Store Injector Properly

*Protect the injector by placing it in a gallon storage bag rather than just putting it into a box.*

*Critical Inspection*

## 9 Inspect Injector Hold-Down

*Look closely and you will notice how the upper bolt of the injector hold-down has fewer threads and more of a shoulder than the lower retaining bolt. Remember this when it is time to reassemble the engine.*

## 10 Remove Intake Manifold

*At this time, you can remove the intake manifold from the driver-side cylinder head.*

## Driver-Side Cylinder Head

### 1 Remove Rocker Arm

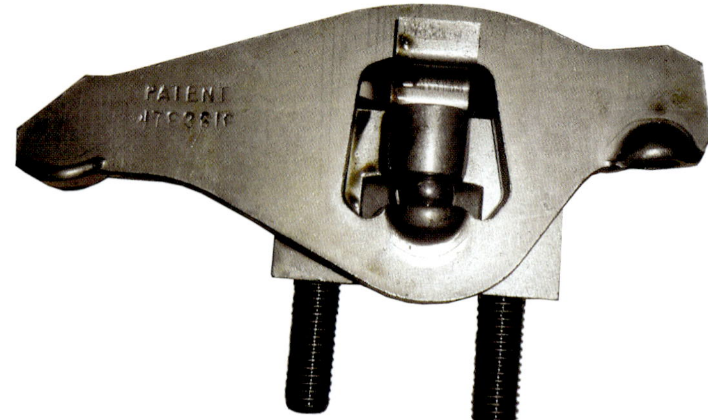


*Using an 8-mm socket, remove the rocker arms from the cylinder heads and the pushrods (left). Use caution when removing the rocker arm. In the lower part of the rocker arm is the pivot ball (right), which can be easily lost.*

### 2 Remove Exhaust Manifolds

*Next, remove the exhaust manifolds from the cylinder head using a 13-mm socket.*

## Coolant System

### 1 Remove Water Pump Pulley Bolts

*Next, let's work on removing the front structure of the engine. Start by removing the bolts that hold the water pump pulley.*

### 2 Remove Return Pipe

*Remove the heater core coolant return pipe from the top of the water pump.*

## 3 Loosen Feed Tube

*Use a 9/16-inch wrench to hold the exhaust backpressure sensor while using a 5/8-inch wrench to loosen the tube that feeds the sensor from the passenger-side exhaust manifold.*

## 4 Remove Manifold Feed Pipe

*Remove the exhaust backpressure feed pipe from the passenger-side manifold with a 5/8-inch wrench.*

## 5 Remove Exhaust Backpressure Sensor

*Using a 1-inch wrench, remove the exhaust backpressure sensor from its bracket.*

## HPOP System

### 1 Remove Window Cover Bolts

*Remove the bolts to the small window cover using an 8-mm socket.*

### 2 Remove Window Cover

*The factory seals this window cover with sealant, so use a small pry bar to remove the cover (left). Now the bolt for the HPOP is exposed (right).*

## 3 Remove Drive Gear Bolt

*Use an 18-mm socket and an impact gun to remove the bolt that holds the drive gear to the HPOP.*

## 4 Remove HPOP Bolts

*Looking at the rear of the HPOP, you will notice two bolts that hold the HPOP to the front cover (left). Using a 10-mm socket and extension, remove the bolts in the rear of the HPOP (right).*

## 5 Remove HPOP

*The HPOP can be removed from the front cover of the engine (left). When the HPOP is removed, oil drains from the HPOP reservoir, which is in the top of the front cover (right).*

## 6 Remove HPOP Reservoir Bolts

*On top of the front cover of the engine is the HPOP reservoir. Remove the bolts from the top and remove the reservoir from the front cover.*

## 7 Remove HPOP Drive Gear

*After removing the reservoir, you can now remove the HPOP drive gear from the front cover.*

# Passenger-Side Cylinder Head

## 1 Remove Dipstick Tube Bracket

*Next, work on the passenger side. Start by removing the engine oil dipstick tube from the bracket on the valve cover.*

## 2 Remove Dipstick Tube

*Once the dipstick tube is free from the bracket on the valve cover, pull the dipstick tube out of the oil pan.*

## 3 Remove Valve Cover

*Now remove the rest of the bolts from the valve cover and remove the valve cover from the cylinder head.*

## 4 Remove Exhaust Manifold

*Using an impact gun and a 13-mm socket, remove the exhaust manifold from the passenger-side cylinder head.*

## 5 Remove Glow Plugs

*At this time, go ahead and remove the wiring harness, injectors, and glow plugs from the passenger-side cylinder head.*

## 6 Remove Cylinder Head Bolts

*Using a 15-mm socket and an impact gun, remove the cylinder head bolts. Then you can remove the cylinder head. Be careful—the cylinder head is very heavy!*

Important!

## 7 Inspect Exhaust and Intake Valves

After removing the cylinder head, take a look at the combustion side. You will notice that on the 7.3 the exhaust valve and intake valve are the same size. The intake valve is always identified by the "dimple" in the center of the valve. Remember this because the valves are not interchangeable.

## 8 Remove Lifter Retaining Plate

*Remove the lifter retaining plate so the lifters can be removed. Using a 13-mm socket, remove the bolts of the retaining plate carefully. Each bolt usually has a shim underneath it, which can be lost easily.*

## Transmission-Related Accessories

### 1 Remove Clutch (Manual Transmission Only)

*Now that the cylinder heads have been removed, use an engine hoist to raise the engine so it can be placed on an engine stand. Start by removing the clutch, if the engine is equipped with a manual transmission.*

### 2 Remove Flywheel

*Next remove the flywheel or flexplate.*

### 3 Remove Adaptor Plate

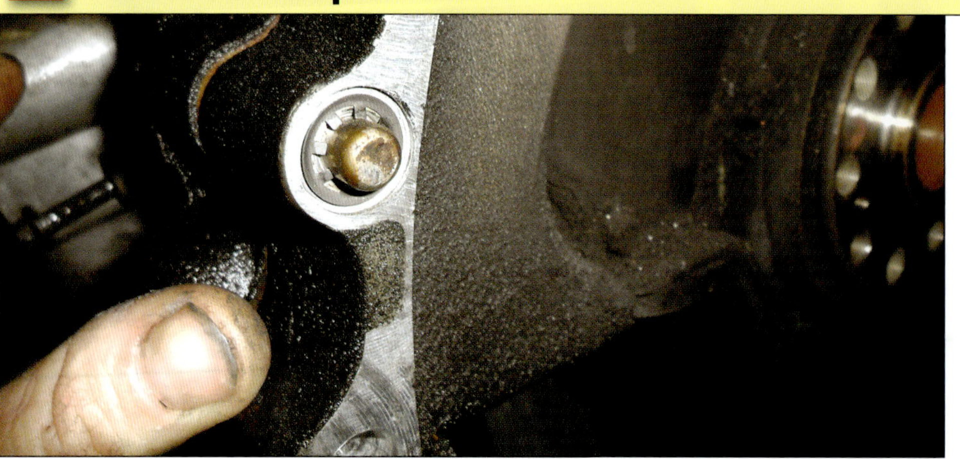

*At the rear of the engine an adaptor plate was used to adapt the International engine to a Ford transmission. The other way around? Go ahead and remove this also by sliding the adaptor plate off the dowels at the rear of the engine.*

## Remaining External Accessories

### 1 Remove Oil Cooler

*Using a 10-mm socket, remove the oil cooler from the engine. Two bolts at the front cover and three bolts at the rear of the block hold it on.*

### 2 Remove Motor Mounts

*Remove the motor mounts from both sides of the engine using a 15-mm socket.*

## Reciprocating Assembly

### 1 Remove Oil Pan

*Now it is time to work on removing the oil pan. There are only twelve bolts that hold the oil pan to the block. The problem is that there is no oil pan gasket, only sealer. This makes the oil pan difficult to remove. Gently pry along the rails of the oil pan the entire length of the block to loosen the sealer so the oil pan does not bend.*

## 2 Remove Oil Pump Pick-up Tube

*When removing the oil pan you notice the oil pump pick-up tube. Remove the two bolts from the oil pump at the front cover and the support nut from the main bearing cap to remove the pick-up tube.*

Documentation Required

## 3 Identify Each Connecting Rod

*Before proceeding further, it is important to mark each connecting rod before it is removed from the cylinder. This identifies which cylinder it came from and also helps place the proper connecting rod cap with the right connecting rod. Each cap and rod are machined together, so if these become mixed up it causes a lot of problems. Use a set of machinist stamps and stamp the rod and cap as it faces the oil pan rail of that cylinder.*

## 4 Remove Connecting Rod Cap

*The nuts of the connecting rod bolts are a 12-point design. Using a 12-point 11/16-inch socket, remove the nuts from the bolts that hold the cap to the connecting rod. Then remove the connecting rod cap.*

## 5 Remove Piston

*Use the wooden handle of a hammer or a rubber mallet to push up on the connecting rod to move the piston out of the bore (left). Be careful not to hit the piston oilers that are placed in the bottom of the bores. This piston (right) had high oil consumption and some smoking. The rings were really stuck in the piston.*

## 6 Remove Harmonic Damper

*Using an impact gun and a 15/16" socket, remove the bolt from the center of the crankshaft that supports the harmonic damper (left). After removing the bolt, you need a special tool to remove the damper (right). Remove the damper to gain access to the engine oil pump.*

## 7 Remove Oil Pump Bolts

*After removing the damper from the crankshaft the engine oil pump is exposed on the front cover (left). The oil pump used on the 7.3 was a "gerotor" style. Remove the four bolts from the oil pump using a 10-mm socket (right).*

Professional Mechanic Tip

## 8 Remove Camshaft Position Sensor

*PRO TIP The camshaft position sensor is located above the engine oil pump on the passenger side of the front cover. Remove the sensor using a 10-mm socket. Be gentle; these sensors have an O-ring that may become hard and seal itself to the front cover. The best advice would be to remove the bolt and try to rotate the sensor inside the front cover by rotating the bracket while gently prying on the sensor.*

## 9 Remove Water Pump

*Remove the bolts from the water pump and remove the water pump from the front cover.*

## 10 Remove Front Cover

*There are only four bolts left holding the front cover, remove them with a 10-mm socket and remove the front cover to expose the camshaft.*

## 11 Remove Bypass Plug

*The plug above the camshaft in the top of the engine block on the driver side is the oil short-circuit bypass plug. The bypass protects the HPOP reservoir during cold starts. Use a 3/8-inch ratchet and extension to remove the plug. Be careful because the plug is spring-loaded. Once the plug is removed, pull out the spring and the check ball inside the block.*

## 12 Remove Camshaft Bolts

*The bolts that hold the cam thrust plate to the block are behind the windows of the camshaft gear. Rotate the engine to expose the cam bolts in order to remove them.*

Important!

## 13 Remove Camshaft

Using a 15-mm socket, the main caps can now be removed from the engine so the crankshaft can be removed (above). Take note because the main caps come numbered from the factory along with an arrow indicating their location and their direction (right).

## 14 Remove Piston Oilers

Next, the piston oilers need to be removed from the block. The oilers are located in the lifter valley of the engine. You need a 10-mm socket to remove the bolts that fasten them to the block.

## 15 Remove Galley Plugs

*The final thing that needs to be done is to remove all of the galley plugs before cleaning. There are two plugs above the camshaft at the front and rear of the engine. The rest of the plugs are at the base of the block at the oil pan on the driver's side of the engine.*

# 6.0-Liter Removal and Disassembly

The 6.0 engine was offered from 2003 to 2007. The 2003 model year is somewhat different from later models, as it was the first year for this engine and, as the year models progressed, so did the changes and updates.

Engine removal can be handled in either of two different ways. Because of the body style change (along with changes in engine design), removal of this engine can be challenging. Some technicians prefer to raise the cab to gain access to the components, while others remove the engine to access the necessary fasteners and components in the traditional fashion. While there are pros and cons to each process, I prefer leaving the cab in place. The biggest reason is so the body is not scratched or damaged in any way. There have been many complaints about damage resulting from lifting the cab, so why take the chance?

***Looking under the hood of the 6.0 can be very intimidating. But once several components are removed, the engine will be revealed.***

In order to remove the engine, you need basic hand tools along with a few specialty tools.

The best advice is to take your time and keep track of components as they are removed from the vehicle. The best strategy is to place the components in the order they are removed—in the bed of the truck or in a separate clean area. This helps when it is time for reinstallation.

Before the engine is removed from the chassis, I try to remove as many external components as possible. This includes the components on the top of the engine down to the intake manifold, along with all accessories on the front of the engine down to the front cover. This way, all bolt-on accessories are removed. Final removal of the engine from the chassis then consists of hoisting the long block up and out of the engine compartment.

This method also helps to shed some weight because just the long block weighs around 900 pounds. All of the accessories have to come off anyway, so when the engine is removed it can be placed on an engine stand (refer to Chapter 2 for more detailed information).

The engine bay of 6.0-equipped trucks is a little more cramped than in those trucks equipped with the 7.3-liter. I promise you that some components will be a challenge to remove, but it can be accomplished.

Before you get started, remove the battery cables and battery, and drain the antifreeze and oil.

## Interior Lights

### 1 Remove Air Cleaner

*Start by removing the air cleaner assembly. Once the filter is removed, remove the intake tube assembly going to the turbo inlet.*

### 2 Remove Reservoir

*Remove the coolant hoses from the reservoir and remove the reservoir from the firewall by loosening two screws on top of the reservoir.*

Professional Mechanic Tip

PRO TIP

## 3 Remove Wiring Harness

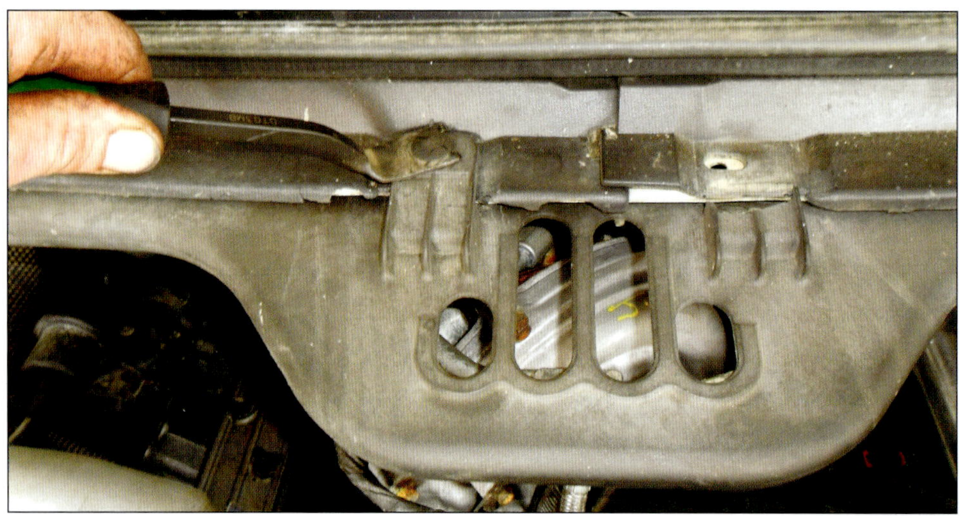

*Remove the wiring harness from under the reservoir bracket and along the cowl of the windshield by removing the "push-in" retainers (left and middle). Fold the plastic shield toward the windshield and secure it to the wiper arms by a zip strap. This gives additional room to access more components.*

## 4 Remove Intercooler Piping

*Remove the intercooler piping from the turbo and both sides of the engine. The clamps can be removed with a 7/16-inch socket.*

## 5 Remove Turbo

*Next, remove the turbo. From the turbine housing, you need to remove the up pipe clamp (1) and the down pipe flange (2) that is a part of the exhaust system. On the 2004 to 2007 models, there is one bolt that supports the turbine housing (3) to the turbo pedestal that needs to be removed. For the 2003 model, the turbo pedestal design is different and this bolt comes in from the side instead of from the top as shown here.*

## 6 Remove Turbo Bolts

*On the driver's side at the rear of the base of the turbo is a bolt that needs to be removed. Also remove the one on the passenger's side.*

## 7 Remove Turbo

*Now the turbo can be removed from the pedestal. Lift the turbo off the oil drain pipe at the base of the pedestal.*

## 8 Remove Pedestal Bolts

*Remove the four bolts that attach the turbo pedestal to the engine block and remove the pedestal.*

Special Tool

## 9 Remove Engine Fan

*Before removing the drive belt, the engine fan must be loosened from the water pump. In order to do this you need a special tool to break the fan nut free. The special tool comes with two wrenches. Take the wrench that looks like a hook and hold the slotted water pump pulley while using the other wrench to loosen the fan nut. Remove the upper radiator hose if you need more room.*

## 10 Remove Drive Belt

*Now the drive belt can be removed. This can be somewhat of a task when using conventional hand tools. This is what the belt tensioner looks like (left). In the end of the tensioner is a slot for a 1/2-inch ratchet. Insert a ratchet in the slot and place a piece of pipe over it for some additional leverage (middle). When pulling up on the ratchet in the tensioner, the drive belt begins to have some slack (bottom). At this time the belt can be removed from the alternator.*

## 11 Disconnect Wiring Harness

*Disconnect the wiring harness from the alternator and remove the three bolts that fasten the alternator to the intake manifold.*

*At the mouth of the intake appears to be some kind of electronic device. This device was incorporated on early models with some having a throttle plate and others being without. The electronic throttle device is there but no throttle plate. Navistar originally used the electronic throttle to close off air coming into the intake manifold so the engine would siphon more exhaust gas coming in from the EGR valve. Later Navistar discontinued the throttle plate and just used the electronic device as a spacer. On 2005 to 2007 models, the intake manifold has a spacer and no electronic throttle device.*

## 12 Remove Fuel Filter Basket Lines

*Remove the fuel supply and return lines from the secondary fuel filter basket.*

### 13 Remove Oil Filter Housing

*At the base of the oil filter housing are four Torx bolts that hold the oil filter housing to the oil cooler. Remove the four bolts so the oil and fuel filter housing can be removed from the engine.*

### 14 Remove Stand Pipe

*Once the oil filter housing is removed, the oil filter stand pipe needs to be removed. Remove the small Torx screw at the rear of the stand pipe and twist. The stand pipe simply lifts off the oil cooler base.*

## Front Drive Accessories and Radiator

### 1 Remove Fan Shroud Bolts

*Inside the inner fan shroud at the base of the front of the intake manifold are two bolts that need to be removed. These bolts hold the inner fan shroud to the intake manifold.*

## 2 Remove Air Conditioning Condenser

*Remove the front grille and the headlight assembly to gain access to the cooling accessories in the front. At this time the air conditioning system needs to be evacuated. Once the refrigerant is evacuated from the air conditioning system, remove the condenser from the front of the vehicle.*

## 3 Remove Transmission Cooling Lines

*If the vehicle is equipped with automatic transmission, remove the transmission cooling lines from the lower part of the radiator.*

## 4 Remove Cooler Bolts

*Remove two bolts at the base of the transmission cooler that hold the cooler to the front of the lower radiator support.*

## 5 Remove Shields

*On each side of the radiator at the top of the upper radiator support are two rubber shields that are held in place with plastic pins. Remove the rubber shields.*

## 6 Remove Intercooler

*In order to remove the intercooler and radiator the upper radiator support has to be removed first.*

## 7 Remove Radiator

*Remove the lower radiator hose and the two bolts in the top of the fan shroud from the radiator. Remove the radiator.*

## 8 Remove Outer Fan Shroud

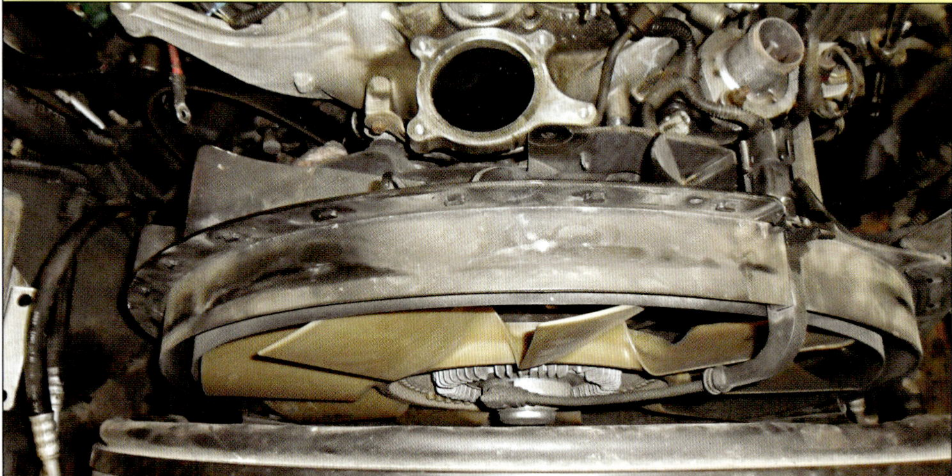

*Remove the outer fan shroud, which exposes the engine fan and the inner fan shroud.*

## 9 Remove Inner Fan Shroud Bolts

*Earlier, the inner fan shroud bolts were removed from the top of the intake manifold. In order to remove the fan and inner fan shroud, two more bolts must be removed from the front of the engine on each side.*

## 10 Remove Inner Fan Shroud

*The fan removes from the water pump pulley in a counter-clockwise rotation. Remove the fan and the inner shroud to expose the front of the engine.*

## 11 Remove Air Conditioning Compressor

*At this time, remove the air conditioning compressor from the engine. This is not easy! The bolts come up through the bottom of the compressor to the engine block. Once the compressor is removed, lay the refrigerant lines to the side and out of the way.*

## 12 Remove Lower Radiator Hose

*Remove the lower radiator hose from the water pump.*

Professional Mechanic Tip

## 13 Remove Power Steering Pump

*To remove the power steering pump, remove the four bolts that hold the pump to the engine block, and set the pump to the side. There is no need to completely remove the pump from the vehicle engine compartment.*

## Remaining Accessories

### 1 Remove Battery Cable Support

*Remove the battery cable support at the lower passenger-side front of the engine block. In this picture, you can get an idea of how the air conditioning compressor is mounted to the engine.*

### 2 Remove Motor Mount Bolts

*Remove the motor mount bolts inside the frame cross-member. There should be two on each side.*

## 3 Disconnect Fuel Lines

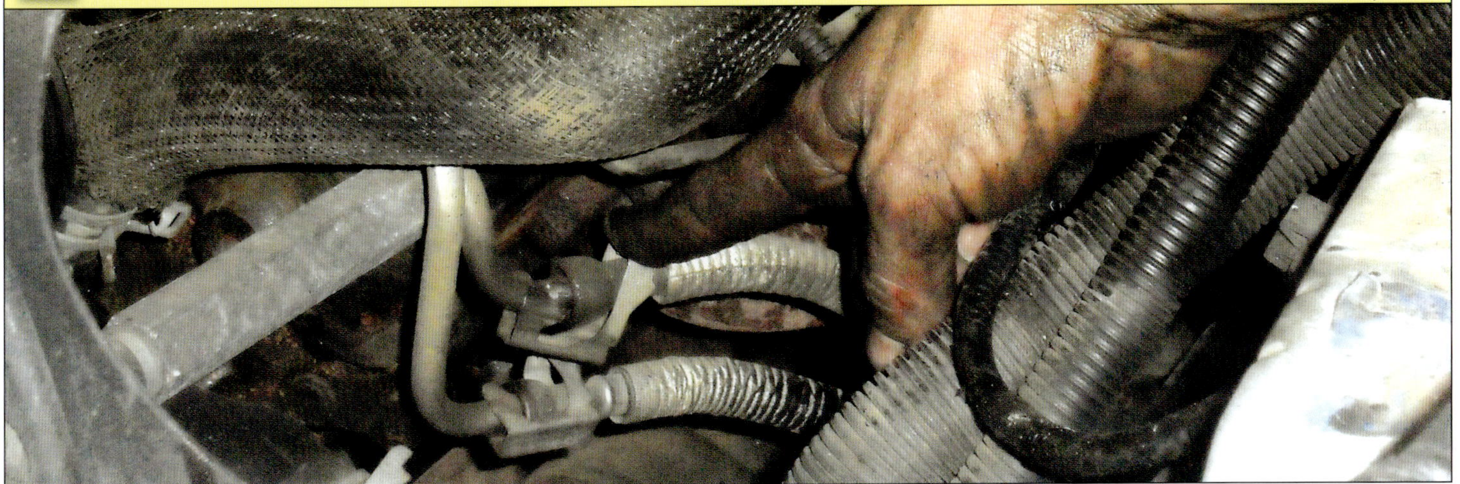

*Disconnect the fuel lines from the engine using a special tool. The tool expands the spring inside the fitting that locks the fuel lines together. Fuel line disconnect tools can also be purchased at your local parts supplier.*

## 4 Remove Starter

*Remove the block heater drop cord at the engine block and the battery cables at the starter. Remove the starter from the engine.*

## 5 Remove Up Pipe Assembly

*Before removing the engine, you need to remove the exhaust up pipe assembly from each exhaust manifold and remove the up pipe assembly from the engine compartment.*

## Remove Transmission (Automatic)

*On the driver's side of the engine just above the oil pan is a window that is covered by a rubber plug in the engine block. This window is used to access the torque converter bolts. Remove the 14-mm torque converter bolts. After the torque converter bolts are removed, remove the transmission bolts also.*

## Remove Engine

### 1 Attach Engine Hoist

*Engine support brackets are located at the front of the driver's side of the engine and at the rear of the passenger's side of the engine. Place a chain tightly through both support brackets and attach to an engine hoist. The engine can now be removed.*

### 2 Remove Engine

*It is very important to remove the accessories from the top of the engine before trying to remove the engine. There is not a lot of room to raise the engine motor mounts from the crossmember, but there is just enough to get the oil pan across the crossmember.*

## Transmission Accessories

### 1 Remove Flywheel

*In order to place the engine on an engine stand, the rear structure of the engine needs to be removed in order to remove the crankshaft. Start by removing the flywheel.*

Important!

## 2 Remove Flywheel Adapter

In order to remove the rear structure of the engine, remove the flywheel adaptor at the rear of the crankshaft. This can be done with a three-finger puller (left). After removing the rear structure, you will notice that there are bolts in the rear of the crankshaft (right). DO NOT remove these bolts! This is the rear seal adaptor that is fitted to the crankshaft. If you remove these bolts and remove the rear seal adaptor, the crankshaft cannot be reused!

# Injectors and Glow Plugs

## 1 Remove Valve Covers

*Now the valve covers can be removed. Make sure to mark the valve cover as to where the bolts and studs are located. I usually scribe the valve covers with small lines.*

## 2 Remove Oil Rails

*Next, the high-pressure oil rails can be removed. This style of rail is for the 2003 model. Notice that the high-pressure oil rail is fed by a braided line (above). The braided line needs to be unlocked and removed in order to remove the oil rail. There is a special tool for this also. If you do not have this special tool, the high-pressure oil line can be removed by using a screw driver to push down on the locking spring (left).*

Special Tool

## 3 Remove Injector Connectors

*Before removing the rocker boxes, you need to remove the injector connectors for each injector that passes through the rocker box. There is a special tool that collapses the locks of the injector connectors. Once the locks of the connectors are compressed by the special tool, push the connector through the rocker box. The special tool (PN 6766) is made by OTC.*

## 4 Remove Injector Hold-Down

*Using a number-40 Torx bit socket, remove the injector hold-down from the cylinder head so the injectors can be removed.*

## 5 Remove Glow Plug Harness

*Remove the glow plug harness from the rocker box.*

## 6 Remove Glow Plugs

*Once the glow plug harness is removed, the glow plugs can be removed from the cylinder head.*

## Cylinder Heads

### 1 Remove Cylinder Head Bolts

*The rocker boxes are held by the cylinder head bolts. Remove the cylinder head bolts so the rocker boxes can be removed. The cylinder head does not come off at this time because there are smaller bolts at the top of the cylinder head that need to be removed first.*

### 2 Remove Exhaust Manifolds

*Remove the exhaust manifolds from the cylinder heads.*

## 3 Remove Cylinder Head

*In the top of the cylinder head between the intake ports are small bolts that fasten the cylinder head to the block. Remove these bolts so the cylinder head can be removed (left). With the cylinder head removed, you can see the amount of damage to the engine. For some unknown reason, the engine dropped both of the exhaust valves causing catastrophic damage (middle and bottom).*

## Disassemble Reciprocating Assembly

### 1 Remove Oil Pan Bolts

*Remove the bolts that support the oil pan to the engine block.*

### 2 Remove Oil Pan

*The upper oil pan is really an adaptor used to fit the engine with a larger oil pan. Because a wider oil pan would be needed for oil capacity, an adaptor is used between the engine block and oil pan (left). Remove the upper oil pan and the oil pump pick-up tube from the bottom of the engine (right).*

**Documentation Required**

### 3 Remove Pistons and Connecting Rods

*Remove the pistons and connecting rods. Make sure to stamp the connecting rod and the cap to the corresponding cylinder it belongs to. At this time also remove the lifters by removing the lifter hold-down assembly. Each lifter hold-down supports four lifters.*

## 4 Remove Bed Plate

*Rotate the engine on the stand so that the crankshaft faces upward. Remove the bed plate, which is the main webbing of the engine that supports the crankshaft and bearings.*

## 5 Prep for Machining

*Now the camshaft and the rest of the accessories such as the high-pressure oil pump, high-pressure oil lines ("branches" in the rear of the engine), and oil cooler can be prepped for machining.*

CHAPTER 4

# Inspection and Machine Shop Work

The Power Stroke is a durable engine designed to last a long time. Most Power Stroke engines that are in need of a rebuild generally have between 350,000 to 400,000 miles. There are a few daily-driven trucks out there with this many miles, but most of these are used for businesses such as construction or towing. A Power Stroke rebuild can be expensive when compared to rebuilding a traditional gasoline engine.

*When rebuilding your Power Stroke engine, take the time to find a machine shop that provides a first-class job. Machining techniques have changed over the years and these newer advances are needed to ensure the engine lasts another 350,000 to 400,000 miles.*

One convenience with this rebuild is being able to use aftermarket parts. Over the years, more companies in the rebuild industry have started offering parts for these 7.3 and 6.0 Power Stroke engines. While these aftermarket companies may offer reputable parts at a less expensive price, the major expense of the rebuild comes from the teardown, cleaning, and machining. The engine pieces are bigger and heavier than their gasoline counterparts.

The final rebuild total often depends on how many new parts need to be purchased. The average price of a rebuild (including engine removal, machining, reassembly, and installation) often starts around $8,000 to $10,000. This does not include rebuilding or replacing the turbocharger, fuel injectors, or any electrical components (such as sensors or actuators) that you may need.

After the engine has been torn down, all parts need to be "hot tanked." This refers to a vat-type cleaning system with strong chemicals that removes all grease, dirt, soot, and debris so proper measuring

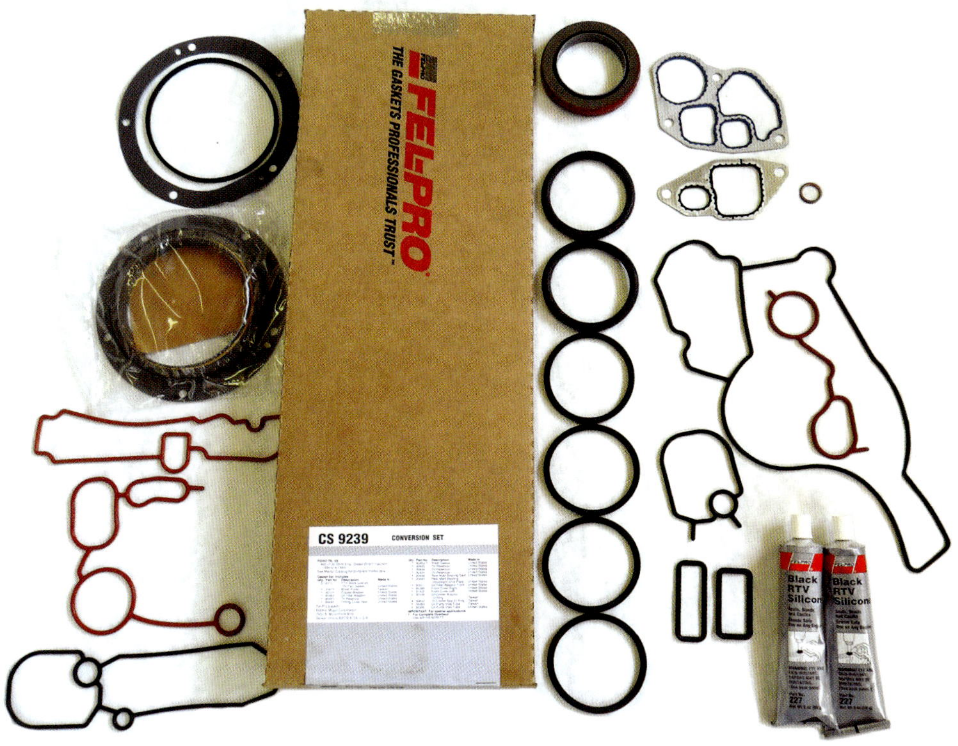


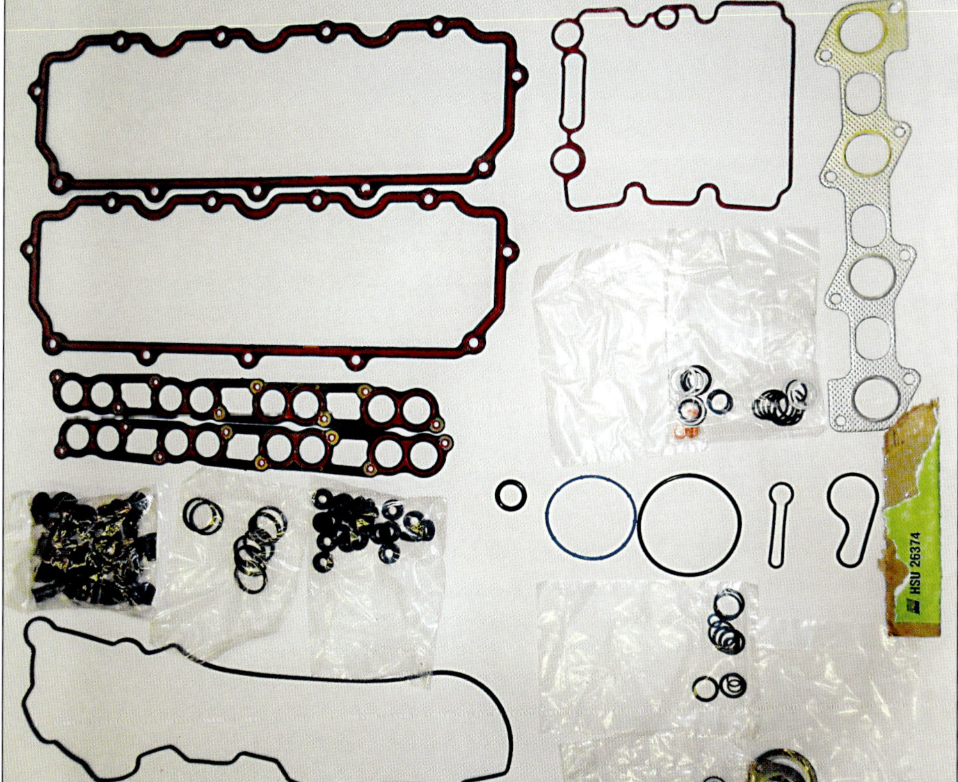


*Fel-Pro gaskets have been the number-one choice by engine builders for sealing an engine. Fel-Pro offers a complete line of gaskets for most Diesel applications.*

and machining can be performed. With the dirt and grease out of the way, a good cleaning of the engine components also indicates what condition the engine is in. This is when you take various measurements to help indicate what machining needs to be performed. You also inspect many parts for cracks that require repair and/or replacement of parts, should damage be found.

In order to perform a first-class rebuild, certain steps and various techniques can be performed while machining to ensure long engine life. Pay close attention to little details such as piston height. On the 7.3 and 6.0 engines, the piston comes out of the bore approximately .030 inch. This is very important in machining, because it affects compression ratio. This is where the proper pistons need to be determined based on how much the bore is worn and whether the deck of the block needs to be resurfaced.

Most engines with considerable mileage need to be overbored in order to "square" the cylinders. Diesel blocks are thick castings, so core shift isn't as much of a problem as bore taper. I find that Diesel blocks often develop bore taper between .004 and .006 inch. The most common overbore for the Power Stroke engine is .020 inch. Most of the time, if the cylinder bores have not suffered any damage, performing a .020-inch overbore is sufficient to bring the cylinder bores back into serviceable condition. This includes a fresh sealing surface for the piston rings and a taper that is within factory specifications.

The block is placed in a boring and milling machine and is squared. For a .020-inch overbore, you have to take into consideration how much piston-to-wall clearance the piston needs. Diesel applications generally

***During the teardown, make sure to measure how far the piston comes out of the bore. This is a contributing factor for choosing the right piston.***

have .0055-inch piston-to-wall clearance. For example, a 7.3 engine has a stock bore measurement of 4.11 inches. If you overbore the cylinders .020 inch the finished bore would be 4.13 inches. The piston measures 4.1245 inches, making the piston-to-wall clearance .0055 inch. When boring the engine, .017 inch is removed from the block with the carbide bits and then the remaining .003 inch is removed by stones in the finish honing process.

While the block is still set up in the milling machine and the overboring process is complete, the block's deck needs to be resurfaced. Most Power Stroke engine blocks typically warp between .004 and .006 inch across the deck. But remember, the pistons came out of the block approximately .030 inch. So in order to machine the deck of the block .010 inch, the piston needs to be shorter, or "destroked" .010 inch.

Mahle Manufacturing offers Diesel pistons in destroked sizes so the deck of the block can be milled. If the pistons were not offered in a "destroked" application, the deck would be milled and the piston would protrude the block .040 inch. This would alter the compression ratio of the engine along with other characteristics of the burn cycle.

When decking the block, several passes are made in order to properly machine the deck surface.

***To begin the machining process, the block is placed in the boring machine and squared along the surface of the four corners of the deck.***

*Mahle/Clevite designs and manufactures pistons for a wide variety of applications. Mahle pistons are the number-one choice for NASCAR engine builders. Mahle offers cast Diesel pistons from high-strength aluminum with a coated skirt made to withstand the pressures and heat from a lower-weight Diesel. The bowl of the piston has been redesigned for maximum efficiency in the combustion process.*

*The surface deck of the cylinder bores need to be cleaned up also to eliminate any warp issues. Make sure to research the appropriate piston choice before decking the block.*

*During the machining process the appropriate surface finish must be acquired. New technology in gasket making has brought about new ways of machining.*

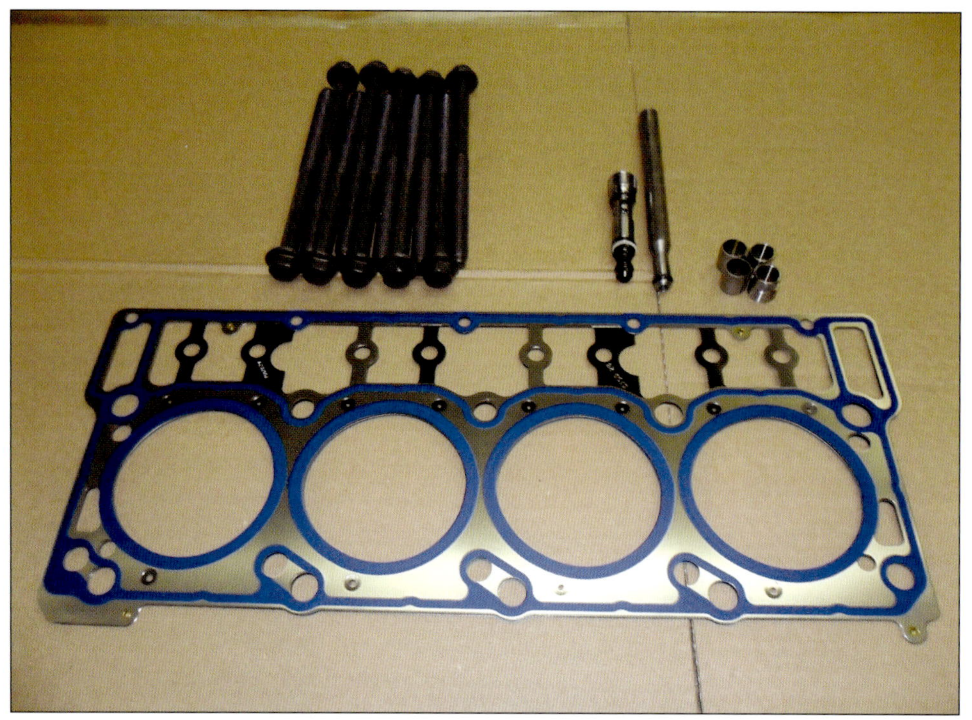

***The 6.0 uses MLS head gaskets. They require certain surface finishes for the most effective sealing.***

After the cylinders have been overbored and the deck has been milled, the block is removed and the cylinder heads are placed into the milling machine and squared. Typical warp in a Diesel cylinder head is also around .004 to .006 inch. For this application, the cylinder head was out around .003. The machining of the cylinder head was the same as the block. For the first pass, .003 was taken off at a slow cutter speed. On the second pass, .001 was taken off and the cutter speed was increased to leave a finish of a 30 Rockwell (Ra).

To finish the cylinder heads, check the valve-to-guide clearance. For Diesel applications, as much as .005-inch valveguide clearance is acceptable.

The valve seats also need to be resurfaced. For the 7.3 and 6.0 engines, the intake valve has a 30-degree seat and the exhaust valve has a 37.5-degree seat. The valves need to be refaced also.

In most Diesel teardowns, the intake valves often take most of the abuse. Replacement of the intake valves is the most common practice. In order to rebuild the cylinder heads properly, the injector cups need to be replaced.

For any rebuild, the crankshaft should always be checked for straightness. Most cranks need to be ground or polished in order to be true. The maximum out-of-round

***The cylinder heads also need to be machined to eliminate any warp issues. The proper surface finish needs to be placed on the surface of the head.***

*The valve job should be performed on the appropriate angle that the manufacturer suggests. Diesel engines do require valve face angles that are different from those of a gasoline engine.*

*In the Diesel, the intake valves seem to take the most abuse. During most rebuilds, the intake valves have suffered too much damage to be refaced.*

*The best piston ring seal comes from the way the block is honed. Piston ring manufacturers often advise the proper bore finish for their piston rings.*

***Because of wear and tear on a Diesel, the crankshaft has some taper on the bearing surfaces. In order to repair the problem, the crankshaft must be machined for straightness.***

limit for the Power Stroke engine is .0002 inch. For most rebuild applications, the crankshaft is usually ground .010 inch. The remaining .0002 is removed by a hand-held polisher with a 600-grit belt. The belt is then changed to 800-grit and polished again for perfection. The rod journals are done the same way.

The connecting rods also need to be machined. This is also called reconditioning. The center of the rod can be resized in a rod resizing machine starting with 320-grit stones to remove the most metal, then using a finer stone of 500-grit to finish.

The little end of the connecting rod utilizes a bushing. The bushings need to be measured also but are usually okay. Most of the time, the bushings can be polished for cleanup without having to be replaced.

If you really want to enhance the internals of your Power Stroke, take the time to have it balanced.

The object of balancing is to match the reciprocating weight, which is the pistons, wrist pin, rings, and the little end of the connecting rod to the reciprocating weight, which is the counterweight of the crankshaft with the big end of the connecting rod with the bearings. The balancing of the engine makes for much smoother operation and longevity of engine components.

***The connecting rods of the engine seem to always distort. In order to repair the connecting rods, they need to be resized.***

*The "little end" of the connecting rod is bushed because of the piston having a floating wrist pin. Most of the time, the bushing can be polished without being replaced, unless the bushing was exposed to extreme heat or negligent oil changes.*

*Balancing of the engine is not required but can bring attention to some areas of the rotating assembly that may need it. The act of balancing the assembly offers smoother acceleration and extended engine life.*

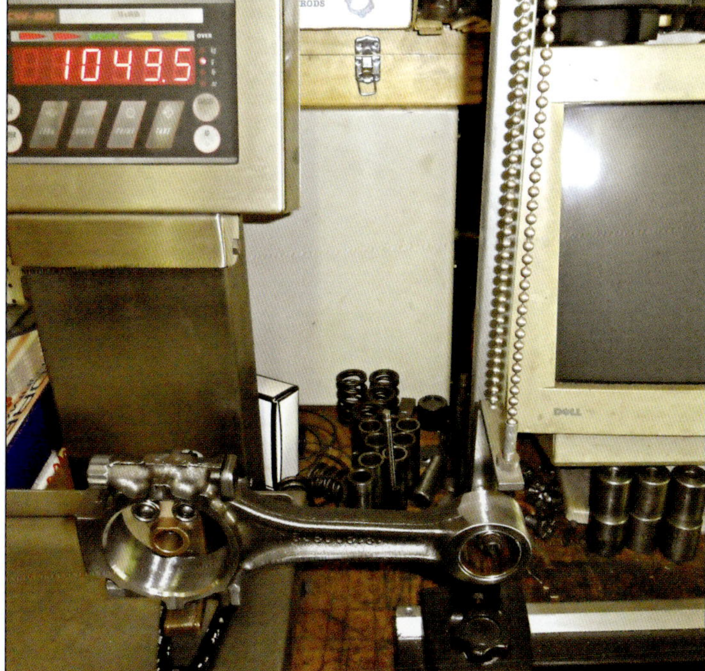


*When balancing the engine, each end of the connecting rod needs to be weighed in order for all the components of the rotating assembly to be matched.*

# Engine Reassembly

After all the machining is performed, the engine can be reassembled in the reverse order that it was torn down.

Of course, I encourage you to follow traditional engine building practices. Clean all sealing surfaces before assembly and installation of any gaskets or sealant, and remember to lubricate all the engine's internal moving parts and bearings during reassembly.

## 1 Paint the Block

***After machining and cleaning the block, mask off all of the internals of the block and paint the block. VHT Paints (a division of Duplicolor) offers a wide range of colors for a variety of applications. The paint is very durable and stands up to the high heat of Diesel components.***

## 2 Reinstall Galley Plugs

***One of the critical things to do at this time is to make sure all of the galley plugs that were removed to clean the block passages are reinstalled.***

## 3 Install Cam Bearings

*At this time, new cam bearings can be installed. When tearing down a Diesel you may notice that all the cam bearings look worn down to the copper layer of the bearing. This is normal and a lot of re-builders choose not to replace the cam bearings unless the engine was exposed to other detrimental issues.*

## 4 Prepare Oil Cooler Assembly

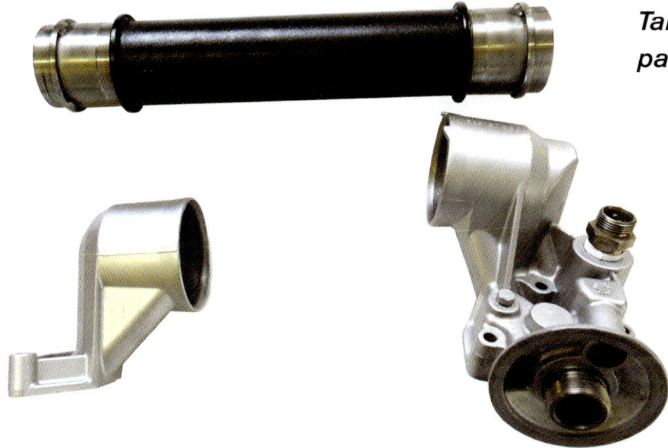

*Take apart, clean, and re-seal the oil cooler. Take the time to paint the oil cooler with VHT paint also.*

# Crankshaft

## 1 Check Bearing Clearances

*In order to perform the proper rebuild, check the bearing clearances. Main bearing clearance should be somewhere between .002 to .0035 inch.*

Torque Fasteners

## 2 Choose Engine Bearings

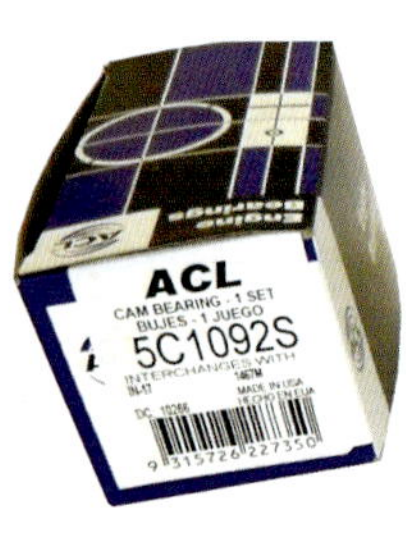

*ACL bearings have provided quality bearings for the OEMs for more than 50 years. For the Power Stroke engine, ACL offers a Duraglide series, which is manufactured of copper, lead, and tin for heavy-duty service and wear.*

## 3 Install Piston Oilers

*Before installing the crankshaft, make sure to install the piston oilers in the main webbing of the block. Apply Loctite 271 to the threads of the bolts to properly secure them.*

## 4 Install Crankshaft

*Install the crankshaft and properly torque the main bearing bolts.*

## Pistons

*These examples give you an idea of how much material on each end of the connecting rod was removed in order for the assembly to be balanced.*

### 1 Install Pistons on Rods

*Lubricate the wrist pins with oil and assemble the pistons to the rods. Make sure to place the piston bowl toward the cam side for the 7.3.*

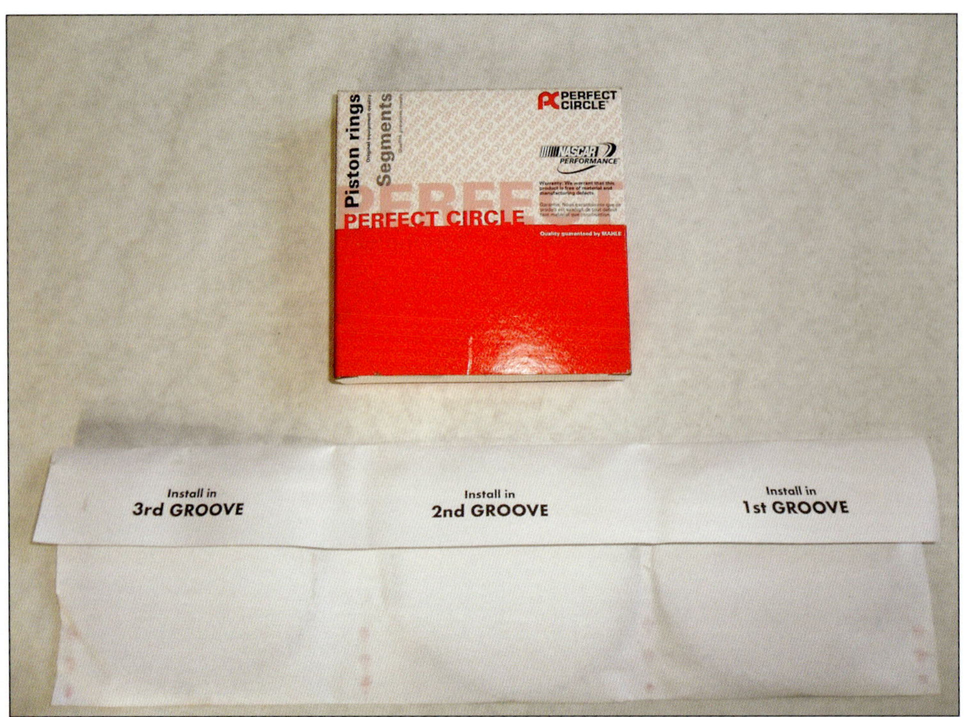


*Mahle pistons are provided with Perfect Circle rings. Perfect Circle has been manufacturing rings for heavy-duty applications for more than 30 years. The top ring is barrel-faced for better seal and quicker break-in, the second ring is a reverse twist for better oil control, and the oil ring is chemically polished for better oil drainback and has a 20-degree gear angle to ensure side seal. All rings are packaged for their correct location on the piston.*

*Performance Tip*

## 2 Inspect Ring Gap

*Before installing the rings onto the pistons, take the time to make sure the rings are properly gapped for the application. Rings are packaged with information regarding the acceptable clearance.*

## 3 Verify Bearing Clearance

*A cheap way to ensure the proper bearing clearance is to use an inside micrometer. This can be used to measure for proper bearing clearance.*

## 4 Install Pistons

*Install the piston rings onto the piston and lightly oil them. Place the rings in the proper position on the piston and install the pistons in the bore with a piston ring squeezer.*

# Camshaft and Lifters

## 1 Install Camshaft

*Install the camshaft into the block and line up the marks on the camshaft with the crankshaft.*

## 2 Install Lifters and Hold-Downs

*After the camshaft is installed and aligned with the crankshaft, the lifters and the lifter hold-downs can be installed in the block. Use Loctite 271 on the threads of the lifter hold-down bolts.*

# Cylinder Heads

## 1 Reassemble Valvesprings

*Luricate the valve stems with oil and insert the valves into the heads. Install new seals and reassemble the valvesprings.*

Save Money 

## 2 Select Head Bolts

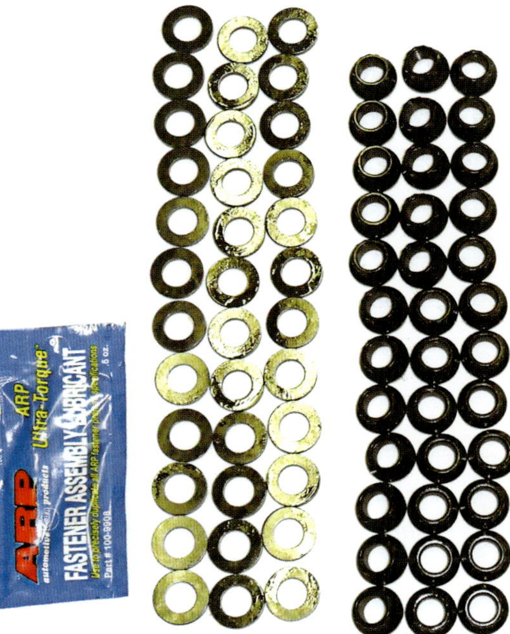

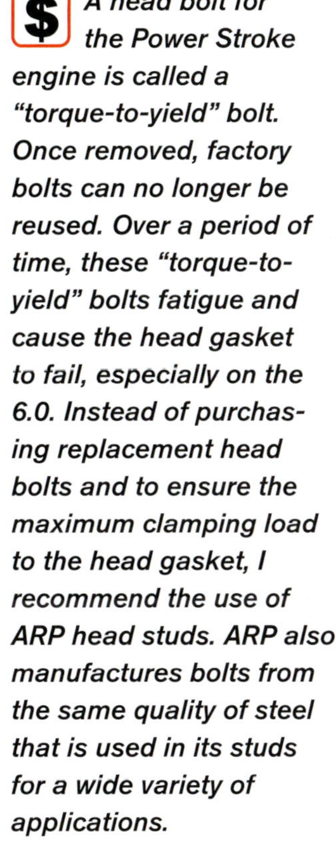

*A head bolt for the Power Stroke engine is called a "torque-to-yield" bolt. Once removed, factory bolts can no longer be reused. Over a period of time, these "torque-to-yield" bolts fatigue and cause the head gasket to fail, especially on the 6.0. Instead of purchasing replacement head bolts and to ensure the maximum clamping load to the head gasket, I recommend the use of ARP head studs. ARP also manufactures bolts from the same quality of steel that is used in its studs for a wide variety of applications.*

## 3 Install Cylinder Heads and Pushrods

*Install the cylinder heads and torque the studs to ARP recommendations. The studs do not torque the same as the factory head bolts. Once the heads are installed and properly torqued, install the pushrods.*

## 4 Install Rocker Arm Stand

*Install the rocker arm stands onto the cylinder heads. Make sure to lubricate the rocker arm ball and pushrod with oil.*

## 5 Install Glow Plugs

*Install the glow plugs into the cylinder heads making sure to coat the threads with Loctite (PN 77164) high-nickel anti-seize lubricant.*

Torque Fasteners 

## 6 Install Plugs into Galley

*At this time, install new plugs into the galley of the front and rear of the cylinder heads for the high-pressure oil passages. These plugs need to be ordered from Ford (PN 4C3Z-6026-CA). Apply Loctite 271 to the threads and torque the plugs to 60 ft-lbs.*

# High-Pressure Oil Pump (HPOP)

## 1 Install Gaskets

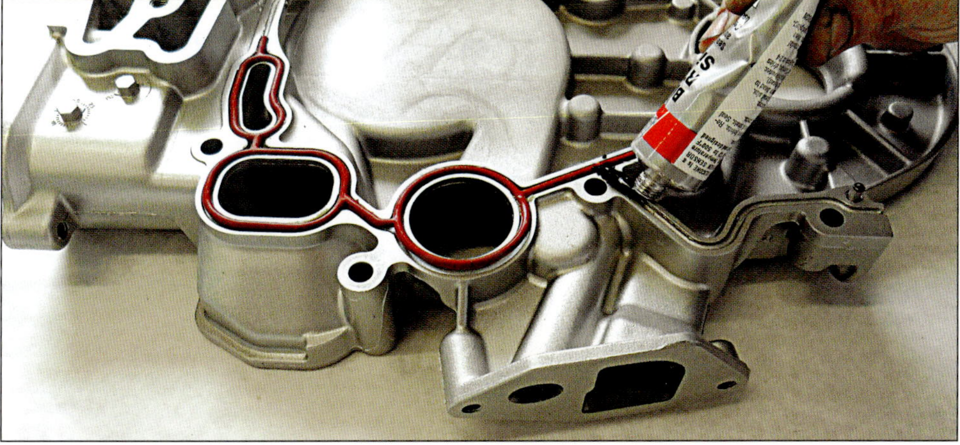

*When installing the new gaskets onto the front cover, you will notice that there is a trough in the front cover that needs to be filled with a small amount of gasket sealer. After applying the sealer, install the front cover onto the front of the engine. The front cover of the engine was previously cleaned and then painted using VHT aluminum spray paint.*

## 2 Insert HPOP Gear

*Insert the gear for the HPOP into the top of the front cover.*

## 3 Install HPOP

*Install the HPOP in the back of the front cover while positioning the gear over the shaft of the pump.*

Torque Fasteners

## 4 Install Drive Gear Bolt

*Install the HPOP drive gear bolt through the front cover and thread into the HPOP. Torque the bolt to 90 ft-lbs.*

## 5 Install Reservoir

*Insert a new gasket into the base of the HPOP reservoir and install the reservoir onto the top of the front cover.*

## Oiling System

### 1 Install Oil Cooler

*After the exhaust manifolds were blasted, they were painted with high-heat cast-iron paint from VHT and secured to the block with new ARP stainless-steel bolts. The oil cooler can be installed at this time also.*

### 2 Install Oil Pump Gear

*When installing a brand new oil pump from Ford, make sure that the word "OUT" is not facing the inside of the front cover as this picture represents. The word "OUT" on the pump needs to face the inside of the oil pump or damage to the front cover occurs.*

## 3 Install Galley Plug

*Make sure the galley plug is installed in the rear of the block beside the camshaft before installing the rear cover.*

## 4 Clean Rear Cover

*Clean the rear cover and apply sealer around the edges of the cover along the bolt holes. This is the only way to seal the rear cover; there is no gasket.*

Special Tool

## 5 Install Rear Seal

*The rear seal needs to be installed. This requires a special tool in order to install it with a wear sleeve. The tool can probably be leased from your nearest parts store. If you choose to purchase the tool from OTC, the price is around $700.*

## 6 Install Oil Pump Pick-Up Tube

*The oil pump pick-up tube can now be installed to the front cover.*

## 7 Install Oil Pump

*Sealed Power makes a replacement oil pump for the 7.3 (PN 224-43626).*

## 8 Install Gaskets

*Gaskets for the 6.0 are notched in relation to were they go. This makes it easy to begin placing the gasket into the component.*

## 9 Install Oil Pan

*Before installing the oil pan, there is a plug on the side of the oil pan that is used to insert the oil indicator tube. Remove the plug and install a new gasket, which consists of an O-ring. This needs to be purchased from Ford (PN 3C3Z-6753-AA). Next, the oil pan can be installed using RTV sealer. This is the only way to seal the pan on the 7.3. An oil pan gasket is only offered for the 6.0 engine.*

### 7.3 Torque Specifications

| | | | |
|---|---|---|---|
| Cylinder head bolts, step 1 | 65 ft-lbs | Exhaust manifold to cylinder head | 45 ft-lbs |
| Cylinder head bolts, step 2 | 85 ft-lbs | Camshaft thrust plate to engine retaining bolts | 50 ft-lbs |
| Cylinder head bolts, step 3 | 105 ft-lbs | | |
| Connecting rod nuts | 70 ft-lbs | Main bearing bolts, step 1 | 75 ft-lbs |
| Engine front cover bolts | 15 ft-lbs | Main bearing bolts, step 2 | 95 ft-lbs |
| Engine rear cover bolts | 15 ft-lbs | Rocker arm pedestal attaching bolts | 20 ft-lbs |

### 6.0 Torque Specifications

| | | | |
|---|---|---|---|
| Cylinder head bolts, step 1 (m14) | 65 ft-lbs | Engine front cover | 18 ft-lbs |
| Cylinder head bolts, step 2 (m14) | 85 ft-lbs | Engine rear cover | 18 ft-lbs |
| Cylinder head bolts, step 3 (m14) | additional 90° | Exhaust manifold | 28 ft-lbs |
| Cylinder head bolts, step 4 (m14) | additional 90° | Camshaft thrust plate | 23 ft-lbs |
| Cylinder head bolts, step 5 (m14) | additional 90° | Main bearing bolts, step 1 | 110 ft-lbs |
| Cylinder head bolts, step 6 (m8) | 18 ft-lbs | Main bearing bolts, step 2 | 130 ft-lbs |
| Cylinder head bolts, step 7 (m8) | 24 ft-lbs | Main bearing bolts, step 3 | 170 ft-lbs |
| Connecting rod (initial) | 33 ft-lbs | Intake manifold | 8 ft-lbs |
| Connecting rod (final) | 50 ft-lbs | High pressure oil rails | 96 in-lbs |

### Standard Torque Specifications

| *Hex Head Thread Diameter* | *Torque ft-lbs* | *Hex Flange Head Thread Diameter* | *Torque ft-lbs* |
|---|---|---|---|
| M6 x 1 | 6 | M6 x 1 | 8 |
| M8 x 1.25 | 15 | M8 x 1.25 | 18 |
| M10 x 1.5 | 30 | M10 x 1.5 | 36 |
| M12 x 1.75 | 51 | M12 x 1.75 | 61 |
| M16 x 2 | 128 | M16 x 2 | 154 |

# Engine Injector Cups

When mentioning the words "injector cups," this often brings about a dazed look from the Diesel owner. Owners of Diesel engines may have heard the term but don't often understand the function.

The cylinder heads of Diesel engines are cast iron or aluminum. The injector itself often has a stainless steel body with rubber O-rings that must seal inside the cylinder head. After a period of years or mileage, the O-rings can distort and moisture can cause pitting of the cylinder head where the O-rings are supposed to seal. When this happens, the cylinder head becomes damaged, which leads to needing a replacement because the O-rings of the injector cannot seal inside the pitted cylinder head.

In order to prevent damaged cylinder heads and to have a good injector seal, manufacturers introduced the injector cup. It is usually made of a soft alloy such as copper so the cup conforms to the cylinder head to seal and also helps cool the injector.

The cup is typically placed in the cylinder head through the coolant passages. Coolant passing through the cylinder head surrounds the cup, helping to cool the injector even more. As an injector is actuated, the moving parts inside it create heat. Removing this heat and keeping the injector cool extends the life of the injector and contributes to more reliable performance.

Do these injector cups ever require replacement? There are only two situations that could justify it. If for some reason the inside surface of the cup (where the O-rings seal the injectors) becomes distorted or destroyed, the cup would have to be replaced to regain the sealing surface. This would be evident in the event of an injector replacement. The second reason would be if Diesel fuel started contaminating the coolant system, and the fuel leak could be traced to a leaking injector cup.

On the Power Stroke engine, the injector seats inside the injector cup. High-pressure fuel surrounds the injector and is sealed by the O-rings. Since the fuel pressure is greater than the coolant pressure, a leaking injector cup causes fuel to seep into the coolant system. Over a period of time, Diesel fuel in the coolant system can cause a lot of damage. Diesel fuel tends to damage anything made of rubber. So after a period of time of being undetected, coolant hoses can blow and water pump seals can be damaged.

If injector cup replacement is necessary, you need special tools. Most of the time, injector cup replacement is

***Often overlooked during the rebuild, injector cups can cause some major problems if they are not replaced. Injector cups are available from rebuilding parts suppliers.***

***A cutaway view of a 7.3 cylinder head shows how the injector cup is placed inside the coolant passage.***

## Engine Injector Cups *CONTINUED*

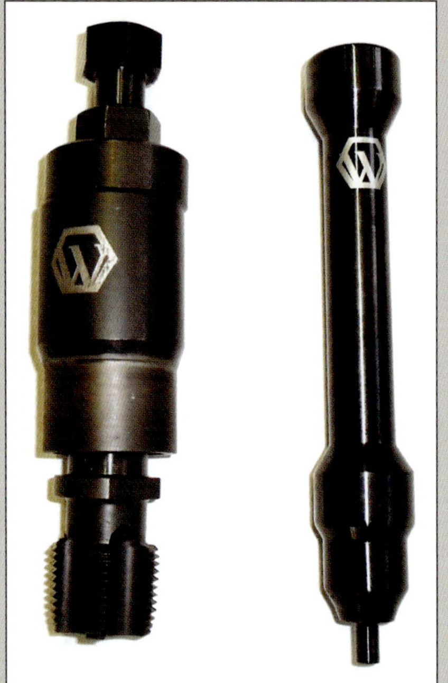

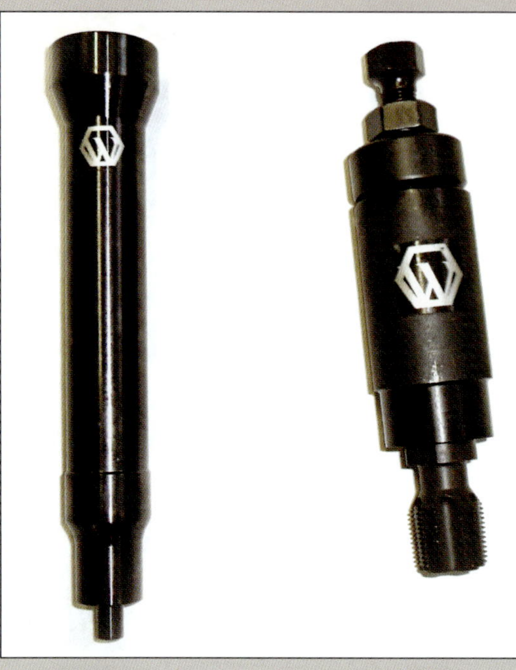

*These injector cup removal (left) and installation (right) tools are manufactured by Whitaker Tools for the 7.3 and 6.0 Power Stroke. Whitaker Tools also offers injector cup tools for other Diesel applications.*

**1** *Place the tap portion of the tool inside the top of the injector cup.*

performed at a dealership because they have the tools required to do the job properly. You can order the tools from Ford to perform the repair yourself, but they are very expensive and it would be difficult to justify the cost for a one-time use. When replacing the injector cups with specialty tools from Ford, the repair generally takes up to 25 hours because the cylinder heads have to be removed.

Whitaker Tools is now making injector cup removal and installation tools for both engines. The great thing about these compact tools is that they were designed to be used in the vehicle without removing the cylinder head.

With the tools from Whitaker, the repair takes about 5 to 6 hours. If the engine runs well, there should be no reason to remove the cylinder head.

If one injector cup is leaking, it probably won't be long before the other cups start to leak also. The only way to determine which injector cup is leaking is to perform a coolant leak check on the cylinder heads. For this test the cylinder heads have to be removed. So by taking about 6 hours for the complete repair using Whitaker Tools, all of the cups can be changed.

If you plan on having someone else rebuild your Diesel, make sure that the injector cups are replaced. They are often overlooked, even when the cylinder heads are machined at a machine shop.

**2** *Rotate the tap portion of the tool in a clockwise direction by turning the hex head at the top of the tool with a 15/16-inch wrench.*

**3** *Once the tap is threaded into the injector cup, rotate the jam nut to the base of the puller and turn in a clockwise direction with a 15/16-inch wrench.*

**4** *This extracts the old injector cup from the cylinder head.*

**5** *Once the cup is removed, the bore of the cup inside the head needs to be cleaned.*

**6** *Use a cross-buff in an air grinder or drill to clean the surface of the bore.*

## Engine Injector Cups *CONTINUED*

**7** *Coat the new injector cup at the top and the bottom with Loctite 609.*

**8** *Install the new cup onto the driver tool. The driver tool is machined just like a real injector. This way the nozzle in the end of the driver is used to align the cup in the cylinder head.*

**9** *With a hammer, strike the injector cup driver. This presses the new injector cup into the cylinder head. Make sure to place the ends of the cylinder head on a piece of wood to support the cylinder head above the table.*

*Installation and removal is now complete and professionally done at a fraction of the cost.*

# Installation and Break-In

After the rebuild, it is time for installation back into the vehicle. The goal is to re-install items as they were removed so as not to have to spend a large amount of time backtracking. That is why I encourage people to lay parts out in the same order as they are removed. It just makes the job easier.

## 7.3-Liter Engine

The 7.3 engine installations are not that complicated. For the 1994 to 1997 models, there is ample room inside the engine compartment and the firewall area to maneuver the engine and install the accessory components.

When working on a 1998 to 2002 model, the front end of the vehicle (hood and fenders) can be removed for easier removal and installation. The installation of the 7.3 engine for this chapter was performed on a 1997-model truck to display some of the difficulty when the components of the vehicle cannot be removed. The Power Stroke Diesel is much heavier than a comparable gasoline engine, so it's important to use the right equipment and proper safety measures when using the engine hoist.

*For 1994 to 1997 models, removal and installation of the engine is a tight squeeze due to the design of the engine compartment. This doesn't pose a real problem; just a little aggravation.*

## Transmission

### 1 Mount Transmission Adaptor Plate

*Mount the transmission adaptor plate to the back of the block. The adaptor is held in place by two dowels. The adaptor needs to be installed before the flywheel or flex plate can be mounted to the crankshaft.*

### 2 Mount Flywheel

*Mount the flywheel or flex plate to the crankshaft. Make sure to use Loctite on the bolts.*

## Engine

### 1 Use Hoist

*Secure a chain through the engine lifting brackets that are fastened to the cylinder heads of the engine. Attach the chain to the engine hoist and lower the engine into the engine compartment.*

## 2 Install Motor Mounts

*After the motor mounts have been placed in the proper location of the crossmember, place the nuts on the studs of the motor mounts and tighten.*

## 3 Remove Hoist

*Once the motor mounts are secure, the engine hoist can be removed and accessories can be installed.*

# Fuel and Oil Systems

## 1 Attach Rear Fuel Lines

*Attach the rear fuel lines loosely to the engine at the fittings on the cylinder heads. When attaching these lines, there is a small rubber washer in each of the fittings. Make sure to purchase and install new rubber washers into the fittings.*

## 2 Install Wiring Harness

*Place the engine wiring harness on the engine. Most of the connections of the sensors can be mated to the wiring harness at this time. (Some connections are installed as the rest of the components are installed.)*

Important!

## 3 Attach Oil Lines

*Attach the high-pressure oil lines to the HPOP and cylinder heads. If the integrity of the high-pressure oil lines is in question, make sure to replace them. This is a very critical component of the Power Stroke engine.*

## 4 Mount Fuel Pump and Filter Basket

*For this 1997 model truck, the fuel pump is mechanical, meaning it is mounted in the center of the top of the engine. The fuel filter basket, along with the mechanical pump, are easier to install if mounted together. The pump and filter basket are attached by a rubber hose at the bottom that are nearly impossible to attach if mounted separately.*

### 5 Mount Turbo

*Mount the turbo to the engine. The turbo is attached to a mounting pedestal and is installed as an assembly. Once the turbo is mounted, the intake tube can be installed to the intake manifold. The rest of the components of the wiring harness can also be completed.*

## Front Drive and Cooling Systems

### 1 Install Accessories

*The accessories of the engine can now be installed. This includes the mounting brackets to the front of the engine, power steering pump, air conditioning compressor, alternator, and drive belt. Do in reverse order from disassembly.*

### 2 Install Radiator

*Install the radiator into the engine compartment. Attach the lower radiator hose.*

Special Tool 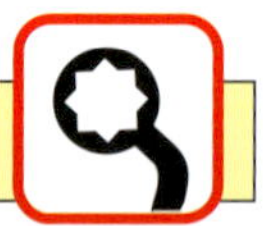

## 3 Install Fan and Shroud

*Install the engine cooling fan and fan shroud between the radiator and engine. Remember, the fan is mounted to the water pump by a large nut* ***cast integrally*** *into the fan. Turn the fan clockwise to fasten to the water pump. Make sure to tighten the fan to the water pump using a special wrench for this application.*

## 4 Install Coolant Hoses

*The rest of the coolant hoses can be installed. This includes the upper radiator hose, heater hoses, and hoses from the degas bottle.*

## 5 Install Turbo Hose

*If your turbo inlet tubing has taken a beating over the years and started to deteriorate (above), there is a replacement turbo hose kit from Ford for this application. The kit (PN F7TZ-9C681-AA) comes with the replacement hose, new mounting hardware, and brackets (right).*

# Exhaust System

## 1 Install Exhaust Up Pipes

*With the engine and components in place on the top side, install the exhaust up pipes from underneath the vehicle using new gaskets. These gaskets come from Ford.*

## 2 Install Exhaust Down Pipe

*Attach the exhaust down pipe to the turbo. If the stock exhaust system is going to be used, the down pipe needs to be placed into the engine compartment before the engine is installed for 1994 to 1997 models. The down pipe for these models does not fit between the firewall and transmission once the engine has been installed.*

## Re-Service Engine

### 1 Service Cooling System

*For the 7.3, reputable green antifreeze can be used (left). Make sure to use a reputable cooling system additive (CSA) also (right). Follow manufacturer directions for your climate conditions.*

### 2 Choose Motor Oil

**Professional Mechanic Tip** PRO TIP

*For break-in purposes and for those of you who use mineral-based oil, I strongly suggest the use of Motorcraft 15w-40 Diesel oil. It was designed specifically for this engine and offers the best in wear protection outside the use of synthetics. The engine holds 15 quarts. Fill the crankcase with 14 quarts and save a quart to be used for the initial startup.*

## 3 Add Oil to Reservoir

*Before startup of the 7.3, remove the small plug in the top right of the HPOP reservoir. Add the remaining quart of oil into the reservoir. This primes the HPOP system for engine startup.*

## 4 Install Fuel Filter

*Make sure to install a new fuel filter into the fuel filter basket.*

### 6.0-Liter Engine

Most trucks with a 6.0 engine use automatic transmissions, and trying to align the torque converter bolts with the flex plate can be a challenge. There are a lot more components on this engine compared to the 7.3, as well as more sensors and a bigger wiring harness. The great thing is that the front end of the vehicle can be removed allowing easier access to components. The installation shown is the way it's typically performed at a shop or repair facility.

## 1 Clean and Paint Frame

*Before installation, be sure to clean the frame area and paint if necessary. Place a jack under the transmission so that the transmission can be raised as high as possible in order to mate with the engine.*

*Professional Mechanic Tip*

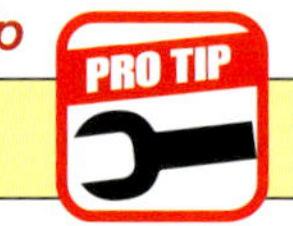


## 2 Install Engine

*PRO TIP Lower the engine into position, mating it with the transmission. Remember, if the transmission is an automatic, make sure to align the studs in the torque converter with the flex plate. To make installation easier, leave the intake manifold off. This allows a little more room for other components to be placed onto the engine.*

## Turbo System

### 1 Install Up Pipes

*After the engine is installed and mated to the transmission, work on placing the exhaust up pipes into the engine compartment. Make sure to install them loosely until the turbo is installed.*

### 2 Install Intake Manifold

*Next, install the intake manifold and properly route the wiring harness so the remaining components can be installed.*

### 3 Install Pedestal and Drain Tube

*Install the turbo-mounting pedestal and the turbo drain tube. Make sure that the turbo oil drain tube is properly seated into the cover of the HPOP with new O-rings.*

## 4 Install Turbo

*At this time, it is best to install the turbo. After the turbo is secure, tighten the exhaust up pipes and the exhaust down pipes.*

# Oil, Fuel and Glow Plug Systems

## 1 Mount Glow Plug Controller

*Mount the glow plug controller and the wiring harness for the injectors.*

## 2 Install Oil and Fuel Filter Housing

*Install the oil and fuel filter housing to the oil cooler and fasten the fuel lines.*

## 3 Install Remaining Components

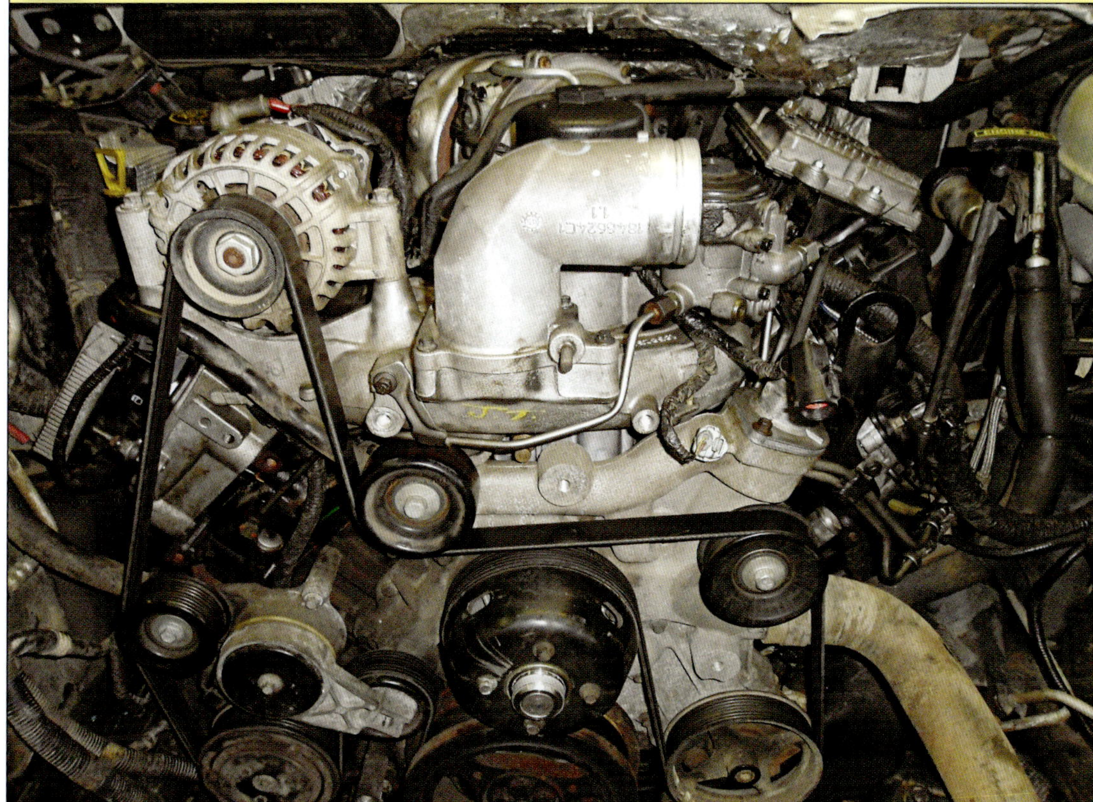

*At this point, the remaining components on the top end of the engine can be installed. These include the fuel injection control module (FICM), alternator, connections of all sensors, power steering pump, air conditioning compressor, drive belt, and fuel line connections on the frame.*

# Finalize Accessory Installations

## 1 Install Inner Fan Shroud

*Next, install the inner fan shroud then attach the clutch fan to the water pump. The inner fan shroud bolts to the intake manifold and engine block.*

## 2 Install Radiator/Transmission Cooler

*Now the radiator and outer fan shroud, intercooler, transmission cooler, and upper radiator support can be installed on the front of the vehicle.*

## 3 Install Battery Tray

*Re-install the driver-side battery tray and connect the engine wiring harness to the PCM.*

## 4 Install Turbo Piping

*Install the intercooler piping from the turbo to the intercooler on the passenger's side and from the intercooler to the snorkel on the intake manifold on the driver's side.*

## 5 Install Headlights and Grille

*The headlights and grille can be reinstalled into the front of the vehicle.*

Important! 

## 6 Install Degas Bottle

*Install the coolant degas bottle, turbo intake tube, and air filter. Be sure to fill the coolant system with the Ford "gold" antifreeze or equivalent for this 6.0 application.*

## 7 Tighten Clutch Fan

*Before starting the engine, remember to tighten the clutch fan onto the water pump using a special tool.*

One thing that needs to be discussed is the subject of pre-oiling the engine before startup. There is really no convenient way of pre-oiling the engine unless you choose to purchase an aftermarket pre-oiling device from an oil pump manufacturer such as Melling. With such a device, you simply remove the oil pressure sending unit and attach the supplied hose in the kit, which feeds oil to the engine's oil galleys from a pressurized tank that you fill with the correct amount of oil.

These devices can be quite costly for a one-time rebuild use. My suggestion is to assemble the engine with proper bearing grease so the bearings are protected during the cranking process. The engine has to spin over quite a few times before there is ignition. All of the high-pressure oil galleys must be filled in order for the injectors to be actuated.

Once it starts, check for leaks. Allow engine to reach normal operating temperature, and drive conservatively for the first 1,500 miles.

## Oil Coolers

For the 6.0 I strongly suggest installing a new oil cooler. New oil coolers for these engines are often overlooked during the rebuild. While some may attempt to clean the cooler, there are often unseen obstructions deep inside that.

While the oil cooler must be purchased from Ford, there are also a few aftermarket companies that offer an upgraded replacement. Whatever the decision, the oil cooler is a major piece of insurance for your newly rebuilt engine.

*Make sure to install the new oil filter screen that is supplied in the oil cooler kit. It filters oil going to the HPOP reservoir.*

*Instructions and all of the necessary components are supplied with a new oil cooler from Ford.*

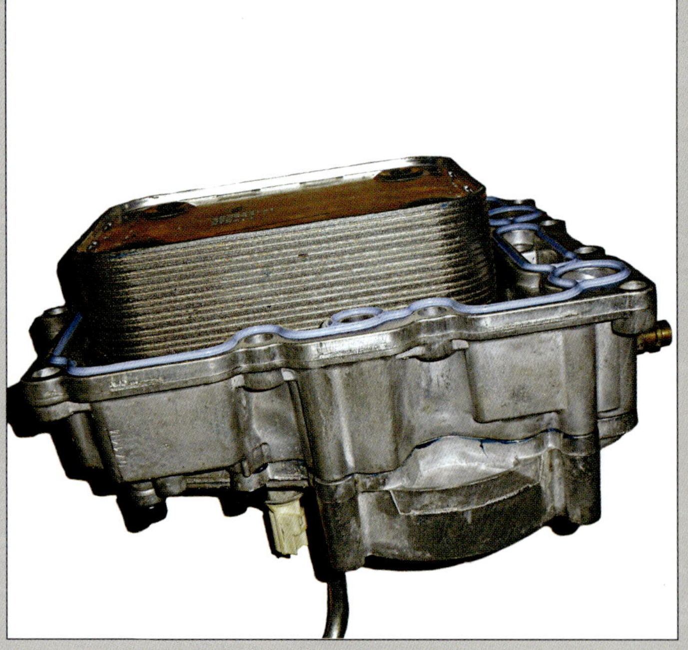

*Install the oil cooler cover after you clean it.*

# Performance Upgrades

In the Diesel world, performance add-ons have grown in popularity as the engines have become more sought after. Over the past decade, truck owners have realized that in order to get more work done the engine must produce lots of torque. Not only does the engine have to produce more torque, but it also must be able to get reasonable fuel mileage. With growing technological advances, not only could the Diesel engine make more power with the same amount of fuel, but it could also be reliable. One reason that people steered away from Diesel engines was because of the noise. Now these engines are very quiet.

There's nothing wrong with obtaining more power. But, for Diesel applications it has to be reliable and still maintain some level of respectable fuel efficiency. If you want to go drag racing or sled pulling, that's different, but most people who purchased a Diesel truck are looking for everyday applications.

Even if your vehicle is a daily driver, there are still plenty of performance upgrades for the 7.3 and 6.0 engines. In order to tune a Diesel, you really have to use a chassis dyno. Even if you purchase reputable parts from a source such as Gale Banks, it is a good idea to find out what the engine makes with and without the add-ons.

When it comes to chasing more power I recommend sticking with the basics. When looking into performance upgrades for Diesel engines, you will find plenty of cold air kits and exhaust systems. Those who take it a step further may install some carefully-chosen gauges on their pillar post. This is a perfect place to start!

*Gary Stecher of Stecher Performance blasts a pass down the dragstrip with his 7.3-powered project truck. This truck is capable of running 11 seconds in the quarter-mile.*

## Cold-Air Kits

The cold-air kit is one of the best basic upgrades available. The colder the air, the denser the air becomes. The denser the air becomes, the more air can enter the engine. The more air that enters the engine, the more power the engine can produce. You may ask why the factory doesn't include free-flowing cold-air kits with these trucks? The fact is that stock airboxes are engineered to minimize noise more than to feed free-flowing air to the engine. If you upgrade to a cold-air kit, you may notice it's a

*This K&N cold-air performance kit was installed on a 1997 F250 with a 7.3. Installation was very easy and neat and also yielded 11 hp more at the rear wheels.*

*Auto Meter offers a complete line of gauges along with a convienent way of mounting known as a pillar pod. The pillar pod is formed to your vehicle's interior pillar post molding. The pillar pod covers and mounts to the existing pillar trim.*

bit louder too. It may also be made of better quality materials than the stock part(s) it replaces. The factory tries to cut costs wherever it can, and this is a good example.

Well, when it comes to cold-air kits, I can only think of one company that has been around for years: K&N Engineering. K&N has been supplying quality gauze filters for more than 40 years. There are several benefits of the gauze element.

First, the gauze element can flow more air than the conventional paper element, so the engine is able to take in more air with the same amount of filter media.

Second, the gauze element never wears out so it can be used over and over. When the filter is ready for cleaning, simply remove it from the vehicle and soak the element with the cleaning solution. Allow the element to air dry, spray an oiling solution into the gauze, and place the element back into the vehicle to be reused again. The cleaning and oiling solution can be purchased from K&N as a filter recharge kit. This ensures proper cleaning and oiling of the filter.

Third, the gauze element can trap smaller particles than the traditional paper element. When should you clean an air filter? Of course this depends on what type of driving conditions the K&N element is exposed to. Under normal highway conditions, K&N recommends cleaning every 50,000 miles.

For Diesel applications, K&N offers replacement OE-style filter elements or cold-air performance systems. Diesel engines are more demanding, so K&N manufactures the air filter elements for Diesel applications with six different types of cotton and deeper pleats in larger sizes. For cold-air performance systems, K&N offers the same air

*This is a K&N cold-air kit installed on a 6.0 in a 2004 F250. For about 20 minutes' worth of work, the gains were impressive on the chassis dynamometer. The truck yielded a 15.5-hp gain and 23.8 ft-lb increase in torque.*

*When installing the exhaust system in 1994 to 1997 Ford F-series trucks, the hardest part is removing the "squished" factory down pipe. The easiest way to remove it is to cut the pipe below the turn, past the turbo flange.*

filter along with OE-style tubing that is designed for your specific application. It takes just a little bit of time with some basic hand tools to transform your vehicle with a cold-air performance package.

Remember, every K&N intake kit has been tested on a dynamometer and proven to increase horsepower. If you purchase a K&N cold-air intake system and you do not see any increase in horsepower on a dynamometer, K&N guarantees to refund your money.

## Exhaust Systems

Similarly, a free-flowing exhaust system helps scavenge the exhaust gases out of the engine, which allows more air to enter the engine.

There are a lot of exhaust companies out there; you just have to know which one to choose. Finding an exhaust system is best approached as a science. There are pressure waves and pulses that take place in the exhaust system of the engine that increase power or take it away. Since there are several exhaust companies that manufacture quality products, you have to narrow the options by price and customer service. I have often found that quite a few companies never return a phone call or email. But, those companies that take the time to communicate are the ones that get my business.

Similar to the cold-air kit, you may find that the aftermarket exhaust systems are a bit louder than the factory setups, and that they may be made from better quality materials (like stainless steel) or they may have an anti-corrosion coating to protect them from rust. The factory simply cannot afford to put exhaust systems like this under the thousands of trucks built every year. But if you're an enthusiast who wants more power, the aftermarket has what you're after.

Flo-Pro has always had great customer service and dyno results. Every Flo-Pro system that I have had the opportunity to dyno-test has always lived up to its reputation. On the Diesel systems that I have tested, there has always been an average of 12- to 15-hp gain and a 25 to 30 ft-lb increase in torque. Along with the additional power gain is the great sound of the performance muffler. It brings the voice of the Power Stroke to the community without being obnoxious.

Flo-Pro was founded in 1981 and has been on the leading edge of exhaust manufacturing. Flo-Pro not only offers an exhaust system that is mandrel-bent but also comes as a direct-fit with no cutting or welding for most applications. Diesel exhaust systems are offered in aluminized steel and 409 stainless steel with your choice of a 4- or 5-inch diameter. Flo-Pro offers exhaust systems for car, truck, and SUV applications. Everything is designed and manufactured in-house, which means great quality, competitive pricing, and customer service.

*The exhaust system from Flo-Pro comes complete with all the necessary hardware and the pipes are pre-bent and fitted for the application.*

*When installing the exhaust system, start by mounting the down pipe to the turbo flange.*

## Gauges

As far as gauges go, they are a great way to see what the engine is actually doing. I often encourage

*The down pipe from Flo-Pro is a two-piece design, which makes it much easier to install. The down pipe is 3 inches in diameter for clearance between the firewall and transmission. After the down pipe, the exhaust system becomes a 4-inch-diameter pipe all the way to the tailpipe.*

people to install the gauges first to get an understanding of what is going on. Install the gauges and drive for several weeks under different conditions to see what changes take place. Then as different parts are added, there should be changes in the gauge readings. Becoming familiar with your gauges can also help in monitoring the health of your engine and/or diagnosing performance issues.

### *Aftermarket Gauges*

An over-the-top gauge is a Diesel air/fuel ratio meter from F.A.S.T. This is something to consider purchasing before you install the exhaust system. The reason is that you need to weld a bung in the turbo down pipe so the air/fuel sensor from the meter can be installed.

The way the meter works is by the sensor reading how much oxygen is in the exhaust system. Diesels run very lean air/fuel mixture ratios. Using this meter is the only way to see what kind of mixture the engine has. Gasoline engines generally run an air/fuel mixture between 12.5:1 and 18:1. (At 18:1, this means that for every 18 parts of air there is 1 part of fuel.)

If you think about it, 12:1 would be a rich air/fuel ratio when compared to 18:1, where the air/fuel ratio is becoming lean. Diesel engines run an air/fuel ratio as high as 80:1. The only time the engine richens is generally at WOT, when the air/fuel ratio is about 15:1.

So if you have plans to install a tuner or programmer into your engine, my advice would be to install an air/fuel ratio meter first. It gives you an idea of how and when the

*All of the exhaust components are made to fit nicely into the existing factory exhaust brackets.*

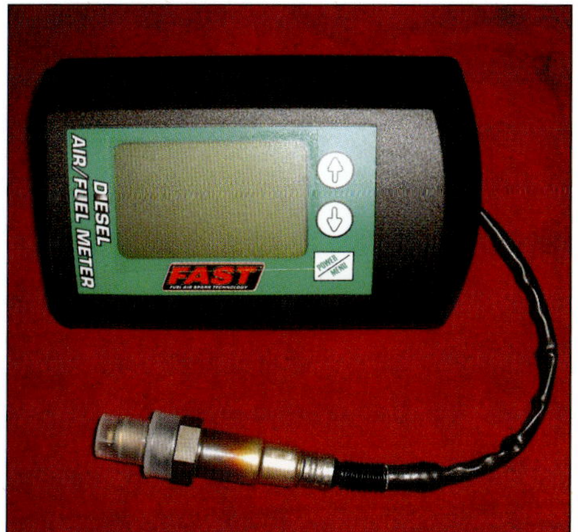


*In order to understand the air/fuel requirements of your Diesel, you have to be able to see them. This air/fuel ratio meter from F.A.S.T., for Diesel applications, is a great tool for tuning.*

*The compression ignition of a Diesel creates unwanted harmonics in the engine's crankshaft. One way to combat the harmonics is by installing a silicon-filled damper by Fluidampr.*

***Auto Meter offers a complete line of gauges along with a convienent way of mounting known as a pillar pod. The pillar pod is formed to your vehicle's interior pillar post molding. The pillar pod covers and mounts to the existing pillar trim.***

engine is making power and how much fuel is needed to obtain this power.

#### *Performance Gauges*

Even before you add performance upgrades, you need to see what the engine is doing. There are certain parameters of the engine that are not monitored in the factory instrument cluster, so you have no idea what is going on with certain things that influence power. The only solution is to install gauges that can monitor other engine functions.

Also, you have to figure out where to put them! Auto Meter has a line of gauges for 7.3 and 6.0 engine applications. Auto Meter also manufactures the pillar pod in which to mount the gauges in the vehicle.

As mentioned earlier, when it comes to aftermarket upgrades you have to go back to the basics. Auto Meter has been building quality components for more than 40 years. Auto Meter has manufactured gauges for all sorts of applications with a variety of sizes and styles. Auto Meter instrumentation has been used in all forms of racing for its accuracy, dependability, and ease of installation.

For a Diesel, the two most desirable gauges measure boost and exhaust gas temperature (EGT). These two gauges are very common, but for the Power Stroke engine, Auto Meter also produces a gauge to monitor the high-pressure oil pump (HPOP). But why should you monitor boost, EGT, and HPOP? Because these three components are vital to engine life and power.

***A boost gauge enables you to see what PSI the turbo is producing. The gauge for this application is mounted in the lowest position of the pillar pod. The reason is, in order for the gauge to read boost, a nylon hose has to connect the gauge to the intake manifold. By placing the gauge in the lowest mounting position of the pod, the nylon hose is easier to route through the dash and firewall to access the engine compartment.***

*Boost Gauge:* Boost refers to the intake manifold pressure that exceeds atmospheric pressure, which is the weight of air that surrounds you. It also depends on your location. The higher the elevation, the less air mass you have so the less pressure you see. (The standard is 14.7 pounds, which is taken from the equation of a column of air in a cross section of one square centimeter from sea level to the top of the earth's atmosphere).

So what is the purpose of the boost gauge? To monitor turbocharger performance. The object of the turbo is to take in air and compress it, and then force it into the intake manifold. This in turn makes air more dense, which forces more air into the cylinders. This is how a turbocharged engine creates more power. The boost gauge tells you how

***A pyrometer gives the operator an idea of just how hot the exhaust becomes under certain conditions. The gauge is a great tool in learning more about your Diesel.***

the turbo is doing and what kind of pressure is needed to make power. As engine modifications are performed, boost from the turbo is influenced.

*EGT Gauge:* This gauge is also known as a pyrometer. Most over-the-road trucks have this gauge in their instrument cluster. Exhaust gas temperature is important to Diesel engines because they are controlled by fuel. The intake manifold of a Diesel engine has no throttle plate to control air. The only way to control air is to adjust the amount of fuel. The more fuel the engine receives, the more power the engine makes, which causes the exhaust gas (as it exits the combustion chamber) to become hotter. There is a point at which the engine will melt down.

When exhaust gas temperature gets high, you have to back off the throttle. So it is a good idea to know how hot the exhaust gets, especially when performing other modifications, to ensure you know when to back off the throttle to avoid engine damage.

The pyrometer is a great tuning and learning tool. Most owners do not realize just how hot the exhaust can get, especially when their engine is in a hard pull. Under normal driving conditions, the EGT generally ranges from 500 to 600 degrees F. When the vehicle is loaded and under a pull, temperatures of 1,000 degrees F or more can be reached.

*HPOP Gauge:* This was a great idea introduced by Auto Meter. As RPM comes up in the Power Stroke engine, so does the pressure from the high-pressure oil pump. This is so higher injection pressures can be reached to obtain more fuel atomization for more power. Not only is the gauge great for monitoring the HPOP, but also to educate the owner of the pressures needed to maintain the engine speed.

This gauge can also be used when tuning with aftermarket software. As more engine power is programmed into the PCM, the greater the demand for high-pressure oil. With the Power Stroke there is a point at which no more power can be achieved because the HPOP cannot keep up. This is very evident if you decide to install bigger injectors. A lot of times when using the gauge, the signs are evident for the need of a bigger HPOP.

## Aftermarket Dampeners

There are products that owners often overlook, but are very important when it comes to making power and for reliability of the engine. For instance, no one ever considers changing the harmonic dampener. People often associate this name with the harmonic balancer, which is slang in the automotive world. It is used by manufacturers as a balancing tool, but its primary function is to dampen harmonics, not balance them. If you want to do your engine a favor and make more power, install a Fluidampr.

*The HPOP gauge is a great addition from Auto Meter. Not only can the gauge be a tool for learning about your engine, but it can also help identify problems in the Power Stroke high-pressure oiling system.*

*The gauges from Auto Meter were made for this Ford truck application. In this 1997 F250, the gauges blended well with the factory instrument cluster.*

*The harmonic damper from Fluidampr is a little wider than the factory version. To keep the fan from hitting the damper, you need to install a fan spacer, which is supplied in the kit.*

*If you are looking for an extreme turbo upgrade for your 7.3, check out this turbo from Turbonetics. This Turbonetics T-series turbo comes with ceramic ball bearings, 76-mm compressor wheel, 68-mm turbine wheel, and supports up to 800 hp and 1,200 ft/lbs at 55 psi of boost.*

Fluidampr is a division of Vibratech Industries manufactured by Horschel Motorsports. Fluidampr is made in the United States and has been around since 1946. What makes the Fluidampr so unique is the silicone gel and internal inertia ring inside the dampener. Stock dampeners are made of rubber and are only effective in a narrow RPM range. Over time, the rubber becomes weaker. As the rubber continues to deteriorate, the less the dampener can absorb. Engine harmonics create a frequency that destroys the engine unless they are absorbed. The more power the engine makes, the more the harmonics change in frequency. Before long, the result is damage to the bearings and crankshaft.

By using a Fluidampr on your engine, especially if the engine has been modified, there can be as much as a 16 hp and 35 ft-lbs increase in torque. That should tell you that harmonics are a big problem in Diesel engines.

## Computer Reprogrammers

A question I am often asked is "Which programmer would you choose to gain power in the Power Stroke engine?" That's a tough one to answer since there are a lot of reputable companies offering tuning modules for these engines. They all are competitively priced and very reliable. It really depends on the specific application and what the truck is going to be used for. It also depends on how many aftermarket parts are already installed on the vehicle. Most companies offer tuning packages that are going to make power, but it is nice to know exactly what changes are taking place to achieve that power. The best choice is to have someone with a chassis dyno with tuning capabilities and experience tune your Diesel engine.

When modifying and making serious changes to the Power Stroke engine, I often rely on Aaron Lail at Tru Dyno Sports. Aaron has been doing performance calibrations with aftermarket EFI and OEM systems for gas and Diesel applications for more than 15 years. He has a vast knowledge of Diesel tuning and offers a variety of custom tuning programs that he has developed using his Mustang 1100SE Eddy current chassis dyno.

## Turbos and Intercoolers

One thing that you may experience is that when you place some

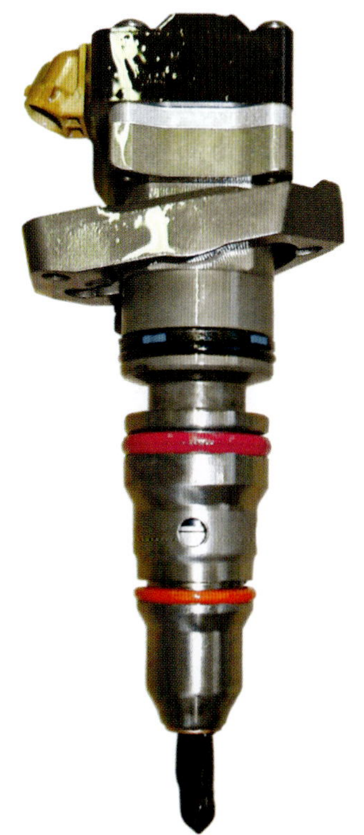

*The injectors we chose from Full Force Diesel in our rebuild of the 1997 7.3 was the Stage 1 upgrade. We mailed them our old injectors and they sent us a set of Stage 1 upgrades, which had been tested, flowed, and were ready to install.*

bolt-ons and programmers on your Power Stroke, you reach a plateau. You wonder what to do next in order to achieve more power. With the Power Stroke, air is not the issue as much as fuel. I can tell you that if the engine is intercooled, you can change intercoolers and turbos to help the air situation a lot.

If you want to change the turbo, stick with an experienced company such as Garrett, which was the original manufacturer of the turbo for the Power Stroke engine. Garrett offers some impressive upgrades. If you are looking for something in the aftermarket, check out turbo packages from other companies such as Turbonetics.

## Injectors and HPOPs

The stock fuel capacity of the injector for the 7.3 engine is 96 cc. There comes a point with an aftermarket programmer where the injector is cycled to its maximum and still can't keep up with the demand.

In order to combat this problem, you have to install a bigger capacity injector. In order to find the appropriate injector, you have to find the right aftermarket manufacturer. The main thing is not to just create a bigger injector but to be able to flow test them so they both flow the same and you have a matching set.

The HEUI design makes for a more elaborate way of testing injectors. Special machines are built to flow-test HEUI designs but are rather expensive. Full Force Diesel has the equipment and years of experience. The company offers different levels of injectors for different applications of the Power Stroke.

Full Force Diesel coats the barrels and plungers of the injectors with tungsten then flow-tests and adjusts them to within 1 percent of one another. Each injector also comes with an 18-month warranty.

Something that you need to keep in mind about the Power Stroke injector is the HPOP. Increasing the injector capacity means that you need something bigger to drive it. Owners often find that when increasing the injector size the HPOP cannot keep up with the demand, and pressure often drops under hard acceleration. This occurs when someone installs a serious programmer with a hot tune. In order to overcome this, the HPOP volume must be increased to keep up with the injector's demand.

Gary Stecher manufactures an HPOP for this specific need. The pump is better known as the "Stealth." Stecher Performance has been developing a better HPOP for the Power Strokes for years. While the Stealth pump may be a little pricey, it is the best one on the market. The Stealth HPOP out-flows any of the competitors' pumps without a pressure drop. This makes for more fuel atomization and power from your Power Stroke engine.

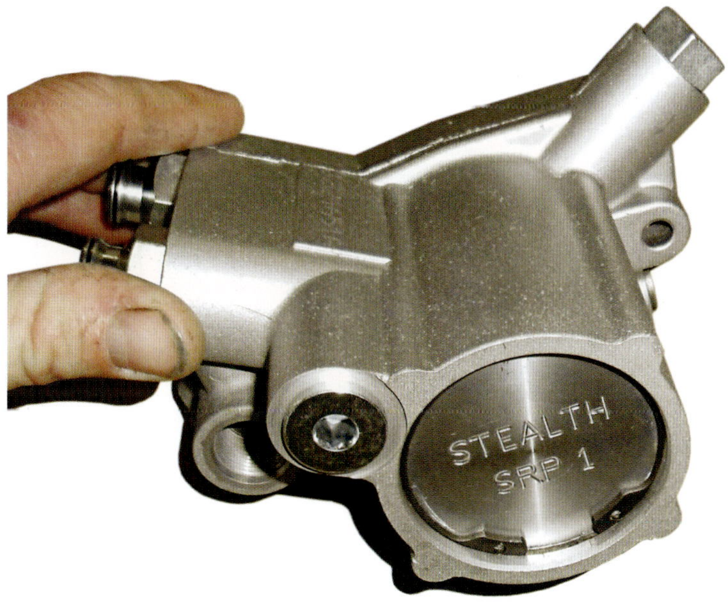


*When performing injector upgrades, you may want to consider upgrading your HPOP. Stecher Performance offers the Stealth HPOP in different stages for various applications.*

*Centerforce offers complete clutch packages to handle the high-torque capacity of Diesel engines. This kit comes with the one-piece flywheel, which replaces the factory two-piece flywheel, along with all the necessary components for a professional installation.*

*The Centerforce clutch is a really nice piece. The small counterweights attached to the pressure plate aid engagement at higher RPM via centrifugal force.*

## Flywheels

With 7.3 trucks, I often find that owners choose the manual transmission. When the Power Stroke Diesel came on the scene, people were skeptical of an automatic transmission behind a vehicle made for towing. Most owners elected to have a manual transmission for more reliability and ease of repair.

Most owners were unaware of the "dual mass flywheel" in a manual transmission. It was designed for the high-compression Diesel engine. It consists of one flywheel centered on top of another flywheel with a friction disc between the two. The flywheels were dampened by a set of springs located on the engine side at the back of the flywheel. The springs were used to dampen a lot of the engine torsional vibrations, which also was easier on the transmission. The use of this flywheel made for more of a cushion and car-like experience in a larger vehicle. The only problem was, after time, the springs in the flywheel wore out.

The flywheel springs wearing out wasn't the problem, it was the replacement price. The price of a replacement "dual-mass flywheel" was around $700, not counting the cost of the rest of the clutch. Owners often found that there were aftermarket companies offering a traditional (one-piece) flywheel design to replace the dual-mass version at half the cost.

The great thing was that the flywheel never needed to be replaced again. Since most of the replacement flywheels are imported, the level of quality is really not what you would expect for performance. The imported versions work great for stock applications. But, if you are going to make more power and change your flywheel and clutch, choose top-quality parts. The best thing to do is go with a complete kit from Centerforce.

Centerforce is a division of Midway Industries, which manufacturers clutch and pressure plate systems for high performance in the United States. Centerforce was started in 1982 by hot rod legend Bill Hays. Bill's design was the Centerforce Weighted Clutch System. This system incorporated a competition-rated clutch and pressure plate with non-asbestos friction facings. The weighted design was developed to increase clamping loads and easy pedal effort.

*The turbo temperature monitor from Isspro mounts easily under the dash and takes just minutes to install. This little box offers cheap insurance in the proper cool down of the turbo.*

Centerforce also carries this same clutch design and a single-mass flywheel for Power Stroke applications. The clutch is rated for 720 ft-lbs of torque and the flywheel is made from billet steel and is SFI approved.

Even though the single-mass flywheel is installed, there have been no reports of transmission damage due to torsional vibrations from the engine. Yes, your Power Stroke engine may make more than 720 ft-lbs of torque but that is at WOT. Under normal driving conditions with power enhancements, this clutch and flywheel kit is more than enough—plus it will last for a long time. Centerforce also offers a clutch and pressure plate for 6.0 applications.

## Turbo Timers

Exhaust heat can be a destroyer of the bearings inside the turbo if the engine has not had time to cool down. The typical driver pulls in somewhere and shuts the engine down without letting it idle. This may be detrimental to the bearings of the turbo as this causes "coking," which is what happens to the bearings of the turbo when the exhaust temperatures are hot and the engine is shut down. The excessive temperature starts frying the oil that is in the bearings, which leaves a nasty film. This film slowly takes its toll on the bearings causing damage to the turbo shaft, which damages the wheels of the turbo. With a Diesel engine, you want to let the engine idle for a few minutes to lower the exhaust temperature. The ideal exhaust temperature for shutdown is around 300 to 350 degrees F.

Sometimes having to let the engine idle to cool down may not be convenient. But, there is a device called a turbo timer that allows the engine to run with the ignition key removed and the doors locked until the optimum temperature of the exhaust is reached. After the safe temperature is reached, the engine shuts down. This is a great form of insurance and is a wise move if you do a lot of heavy-duty work or racing with your turbo Diesel.

# APPENDIX

Cylinder Head Torque Sequence
Intake Side
15
13
11
12
14
9
5
1
3
7
10
6
2
4
8
Exhaust Side

Flywheel Torque Sequence
1
2
7
5
9
10
6
8
3
4

Intake Manifold Fastener Torque Sequence
1
3
5
7
2
4
6
8
16
14
12
10
15
13
11
9

Cylinder Location
L
FRONT
R
2
1
4
3
6
5
8
7

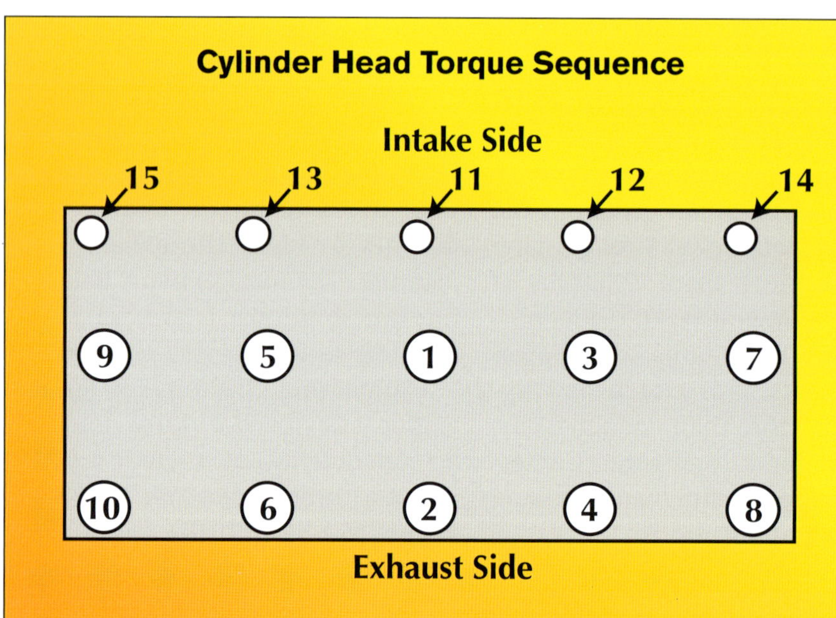

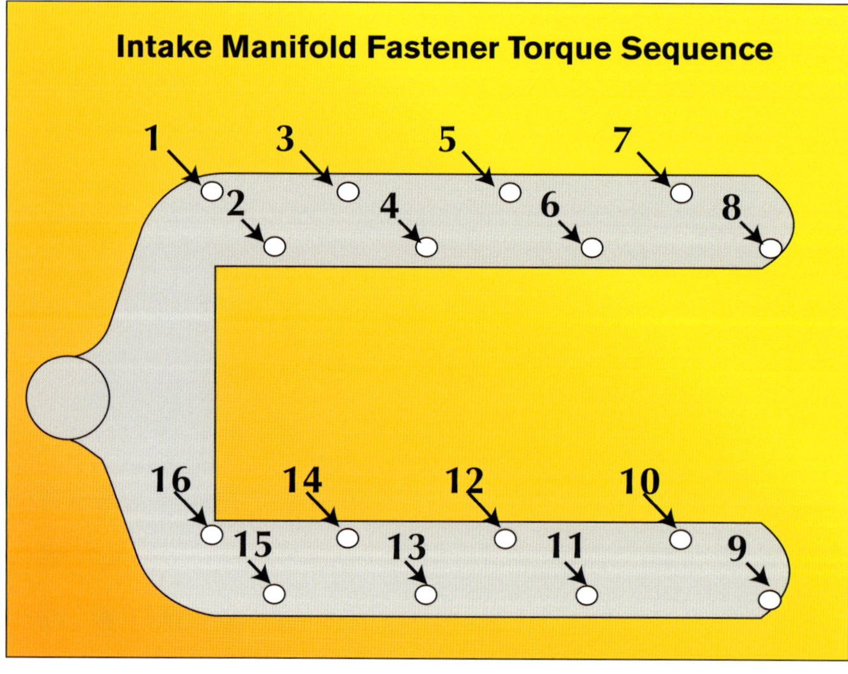

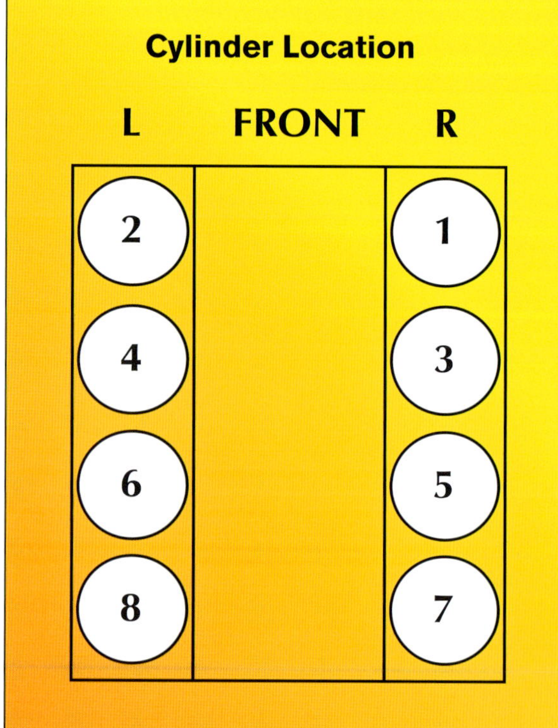

## Basic Specifications

| | *6.0* | *7.3* | *7.3 Turbo* | *7.3 Power Stroke* | *6.0 Power Stroke* |
|---|---|---|---|---|---|
| **Bore (inches)** | 4 | 4.11 | 4.11 | 4.11 | 3.74 |
| **Stroke (inches)** | 4.18 | 4.18 | 4.18 | 4.18 | 4.13 |
| **Displacement (ci)** | 420 | 444 | 444 | 444 | 365 |
| **Horsepower** | 170@3300 rpm | 185@3000 rpm | 190@3000 rpm | 275@2800 rpm | 325@3300 rpm |
| **Torque (ft-lbs)** | 338@1400 rpm | 360@1400 rpm | 388@1400 rpm | 525@1600 rpm | 560@2000 rpm |
| **Compression** | 20.7:1 | 21.5:1 | 21.5:1 | 17.5:1 | 18.1:1 |

## Power Stroke Capacities

| | *Engine Oil* | *Cooling System* |
|---|---|---|
| **7.3 liter** | 14 quarts | 5.75 gallons<br>50/50 antifreeze-water mix |
| **6.0 liter** | 15 quarts | 6.75 gallons<br>50/50 antifreeze-water mix |

## Fault Codes

| *Code* | *Code Description* | *Probable Cause* |
|---|---|---|
| P0046 | Turbo boost control circuit performance | VGT actuator circuit or PCM |
| P0069 | Map/Baro Correlation | VGT, MAP, BARO, or EGR sensors, possible wiring |
| P0096 | IAT 2 sensor performance | IAT 2 sensor, wiring or PCM |
| P0097 | IAT 2 low input | Signal circuit shorted to ground or bad sensor |
| P0098 | IAT 2 high input | Signal circuit open or short to power or bad sensor |
| P0101 | MAF circuit performance | Blocked or obstructed supply to MAF, bad MAF, or PCM |
| P0102 | MAF circuit low input | Signal circuit open to PCM or bad MAF |
| P0103 | MAF circuit high input | Signal circuit to PCM shorted or bad MAF |
| P0107 | MAP/BARO low input | Circuit is open or shorted to ground |
| P0108 | MAP/BARO high input | Circuit is shorted to power |
| P0112 | IAT circuit low input | Circuit short to ground |
| P0113 | IAT circuit high input | Circuit open or short to power |
| P0117 | ECT circuit low input | Circuit is shorted to ground |
| P0118 | ECT circuit high input | Circuit is open or short to power |
| P0196 | EOT circuit performance | EOT showing oil cold to long, bad sensor, wiring, PCM |
| P0197 | EOT circuit low input | Circuit shorted to ground |
| P0198 | EOT circuit high input | Circuit open or short to power |
| P0219 | Engine overspeed | Improper downshift, CKP or CMP interference, PCM |

## Fault Codes *CONTINUED*

| *Code* | *Code Description* | *Probable Cause* |
|---|---|---|
| P0230 | Fuel pump primary circuit | Signal circuit open, bad fuel pump relay, or PCM |
| P0231 | Fuel pump secondary circuit low | signal circuit shorted or open, relay, inertia switch, fuel pump |
| P0232 | Fuel pump secondary circuit high | Circuit shorted to power or sticking relay |
| P0236 | Turbo boost sensor A circuit performance | MAP sensor plugged or bad MAP |
| P0237 | Turbo boost sensor A circuit low | MAP signal circuit short to ground or open, bad MAP |
| P0238 | Turbo boost sensor A circuit high | MAP signal circuit short to power, bad MAP |
| P0261 | Cylinder #1 Injector circuit low | |
| P0262 | Cylinder #1 Injector circuit high | |
| P0263 | Cylinder #1 contribution/balance | |
| P0264 | Cylinder #2 Injector circuit low | |
| P0265 | Cylinder #2 Injector circuit high | |
| P0266 | Cylinder #2 contribution/balance | |
| P0267 | Cylinder #3 Injector circuit low | |
| P0268 | Cylinder #3 Injector circuit high | |
| P0269 | Cylinder #3 contribution/balance | |
| P0270 | Cylinder #4 Injector circuit low | The FICM has detected an open injector circuit or defective coil |
| P0271 | Cylinder #4 Injector circuit high | The FICM has detected an injector circuit shorted to ground or defective coil |
| P0272 | Cylinder #4 contribution/balance | Code set when injectors are not giving the necessary volume of fuel |
| P0273 | Cylinder #5 Injector circuit low | |
| P0274 | Cylinder #5 Injector circuit high | |
| P0275 | Cylinder #5 contribution/balance | Cylinder may be low on compression due to mechanical problem |
| P0276 | Cylinder #6 Injector circuit low | |
| P0277 | Cylinder #6 Injector circuit high | |
| P0278 | Cylinder #6 contribution/balance | |
| P0279 | Cylinder #7 Injector circuit low | |
| P0280 | Cylinder #7 Injector circuit high | |
| P0281 | Cylinder #7 contribution/balance | |
| P0282 | Cylinder #8 Injector circuit low | |
| P0283 | Cylinder #8 Injector circuit high | |
| P0284 | Cylinder #8 contribution/balance | |
| P0341 | CMP sensor circuit A performance | Poor connection or bad sensor |
| P0401 | EGR flow insufficient | EGR valve sticking or stuck, bad EGR valve, EBP sensor |
| P0402 | EGR flow excessive | EGR valve sticking or stuck, bad EGR valve, EBP sensor |
| P0403 | EGR control circuit | Signal circuit open, short to ground or power |
| P0404 | EGR control circuit performance | Bad EGR valve or PCM |
| P0405 | EGR circuit A low | EGR signal circuit short to ground or open, bad EGR |
| P0406 | EGR circuit A high | EGR signal circuit short to power, bad EGR |
| P0470 | Exhaust backpressure sensor | Bad EBP sensor or PCM |
| P0471 | EBP sensor performance | Bad EBP sensor, VGT, or PCM |
| P0472 | EBP sensor low input | Signal circuit short to ground or open, bad EBP sensor |
| P0473 | EBP sensor high input | Signal cicuit short to power, bad EBP sensor |
| P0478 | EBP control valve high input | EBP sensor, VGT valve, stuck VGT, PCM |
| P0562 | System voltage low | PCM voltage is less than 7 volts, bad battery or connection |
| P0603 | PCM keep alive memory error (KAM) | Battery discharged or disconnected, wiring, PCM |
| P0605 | PCM read only memory error (ROM) | PCM failure |

## Fault Codes *CONTINUED*

| Code | Code Description | Probable Cause |
|---|---|---|
| P0611 | FICM performance | Internal FICM failure |
| | | Loss of FICM power |
| P0670 | Glow plug module control circuit | Open or grounded circuit, bad PCM or GPCM |
| P0671 | Cylinder #1 Glow plug failure | Possible wiring, bad glow plug or bad GPCM |
| P0672 | Cylinder #2 Glow plug failure | Possible wiring, bad glow plug or bad GPCM |
| P0673 | Cylinder #3 Glow plug failure | Possible wiring, bad glow plug or bad GPCM |
| P0674 | Cylinder #4 Glow plug failure | Possible wiring, bad glow plug or bad GPCM |
| P0675 | Cylinder #5 Glow plug failure | Possible wiring, bad glow plug or bad GPCM |
| P0676 | Cylinder #6 Glow plug failure | Possible wiring, bad glow plug or bad GPCM |
| P0677 | Cylinder #7 Glow plug failure | Possible wiring, bad glow plug or bad GPCM |
| P0678 | Cylinder #8 Glow plug failure | Possible wiring, bad glow plug or bad GPCM |
| P0683 | GPCM to PCM communication | Possible wiring, bad GPCM, or bad PCM |
| P0684 | GPCM to PCM communication performance | Noise detected in communication line to PCM |
| P1184 | EOT sensor out of range | Leaking thermostat, EOT sensor, wiring |
| P1283 | IPR circuit failure | Open/grounded circuit, stuck IPR, loose connection |
| P1284 | ICP failure aborts KOER CCT test | See codes P1280, P1281, P1282, P1283, P1211 |
| P1291 | High side #1 (right) short to grd. or B+ | Short circuit, faulty IDM |
| P1292 | High side #2 (left) short to grd. or B+ | Short circuit, faulty IDM |
| P1293 | High side open bank #1 (right) | Open circuit, faulty IDM |
| P1294 | High side open bank #2 (left) | Open circuit, faulty IDM |
| P1295 | Multiple faults, bank #1 (right) | Miswired connector or harness, short to ground |
| P1296 | Multiple faults, bank #2 (left) | Miswired connector or harness, short to ground |
| P1297 | High sides shorted together | Shorted wires, faulty IDM |
| P1298 | IDM failure | Internal IDM failure |
| P1316 | Injector circuit/IDM codes detected | Injector circuit failure/IDM codes detected |
| P1378 | FICM supply voltage low | Low batteries, wiring, defective relay |
| | | FICM detects voltage of less than 7 volts |
| | | Low batteries, loose connection, or defective relay |
| P1379 | FICM supply voltage high | Charging system fault |
| | | FICM detects accessive voltage of 16 volts or more |
| | | Most likely charging system fault |
| P1391 | Glow plug circuit low input, bank #1 (right) | Open/short/miswired circuit, faulty relay, glow plugs |
| P1393 | Glow plug circuit low input, bank #2 (left) | Open/short/miswired circuit, faulty relay, glow plugs |
| P1395 | Glow plug monitor fault, bank #1 | One or more glow plugs failed or circuit fault |
| P1396 | Glow plug monitor fault, bank #2 | One or more glow plugs failed or circuit fault |
| P1397 | System voltage out of self test range | Voltage too high or low for glow plug monitor test |
| P2262 | Turbo boost not detected (mechanical) | MAP hose or sensor, intercooler hose leaks, EBP sensor |
| P2263 | Turbo system performance | MAP hose or sensor, intercooler hose leaks, EBP sensor |
| P2269 | Water in fuel | Drain water seperator, bad sensor, wiring |
| P2284 | ICP sensor performance | ICP sensor, IPR sensor, internal HPOP, wiring |
| P2285 | ICP sensor low input | Signal circuit short to ground or open, bad sensor |
| P2286 | ICP sensor high input | Signal circuit short to battery voltage, bad sensor |
| P2288 | Injection control pressure high | ICP sensor, IPR sensor, internal HPOP, wiring |
| P2289 | Injection control pressure high, engine off | ICP ground circuit open, bad ICP sensor |
| P2290 | Injection control pressure low | ICP sensor, IPR sensor, internal HPOP, wiring |
| P2291 | Injection control pressure low, engine crank | ICP sensor, IPR sensor, internal HPOP, wiring |
| P2614 | CMP sensor output circuit | Poor connection from PCM to CMP |
| | | Check Crank and Cam position sensor circuits |
| P2617 | Crankshaft position sensor circuit open | Check Crank and Cam position sensor circuits |
| | | Poor connection from PCM to CKP sensor |
| P2623 | IPR circuit | Open or grounded wiring, bad IPR |

ACL Distribution Inc.
4722 Danvers Dr SE
Grand Rapids, MI 49512
800-847-5521
www.aclus.com

Amsoil
925 Tower Ave
Superior, WI 54880
715-399-8324
www.amsoil.com

ARP
1863 Eastman Ave
Ventura, CA 93003
800-826-3045
www.arp-bolts.com
Engine Fasteners

Auto Meter Products Inc
413 W Elm St
Sycamore, IL 60178
815-895-8141
www.autometer.com

Bulletproof Diesel
134 East Broadway Rd
Mesa, AZ 85210
480-247-2331
www.bulletproofdiesel.com

Centerforce Clutch-
Midway Industries Inc.
2266 Crosswind Dr
Prescott, AZ 86301
928-771-8422
www.centerforce.com

Comp Cams
3406 Democrat Rd
Memphis, TN 38118
800-999-0853
www.compcams.com
Engine Analyzers

Isspro
2515 NE Riverside Way
Portland, OR 97211
503-288-4488
www.isspro.com

Federal-Mogul Corporation
26555 Northwestern Hwy
Southfield, MI 48033
800-325-8886
www.federalmogul.com

Fluidampr
180 Zoar Valley Rd
Springville, NY 14141
716-592-1000
www.fluidampr.com

Full Force Diesel
10696 Franklin Rd
Murfreesboro, TN 37128
615-962-8291
www.fullforcediesel.com

Hickok Inc.
10514 DuPont Ave
Cleveland, OH 44108
216-541-8060
www.hickok-inc.com

K & N Engineering Inc
1455 Citrus St
Riverside, CA 92507
800-858-3333
www.knfilters.com

Loctite Henkel Corporation
26235 First St
Westlake, OH 44145
800-624-7767
www.loctite.com

Mahle-Clevite
www.mahle.com

Pencool-Penray
1801 Estes Ave
Elk Grove Village, IL 60007
800-322-2143
www.penray.com

Stealth Engineering
1375 Oakridge Estates Dr
St.Clair, MO 63077
314-766-7867
www.stealthpumps.com

Turbonetics Inc.
2255 Agate Ct
Simi Valley, CA 93065
805-581-0333
www.turboneticsinc.com

VHT Paints-Duplicolor
101 W Prospect Ave.
Cleveland, OH 44115
www.vhtpaint.com

Whitaker Tools
P.O. Box 175
Mannsville, KY 42758
270-495-5681
www.whitakertools.com

# Additional books that may interest you...

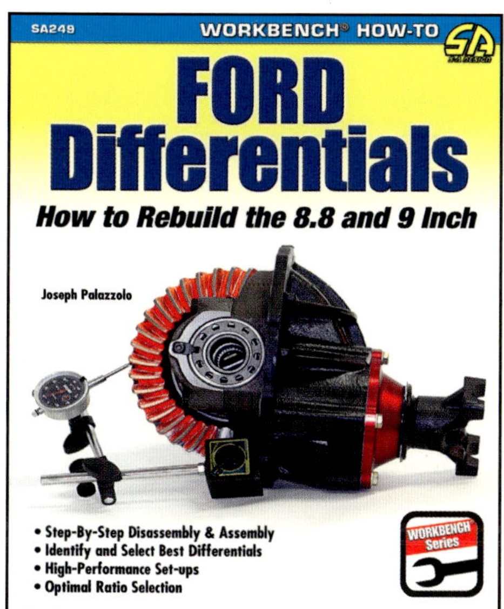


**FORD DIFFERENTIALS: How to Rebuild the 8.8 and 9 Inch** *by Joe Palazzolo* The 8.8-inch Ford differential is popular because of the many Mustangs on the road and being modified today; the 9-inch differential is popular because of the performance Fords from the 1960s and 1970s, and their use for racing and swaps into other chassis as well. This book covers everything you need to know, in step-by-step fashion. Whether it's a straight rebuild or you are swapping performance gears into your car, this guide has it covered. Softbound, 8.5 x 11 inches, 144 pages, 400 color photos. ***Item # SA249***

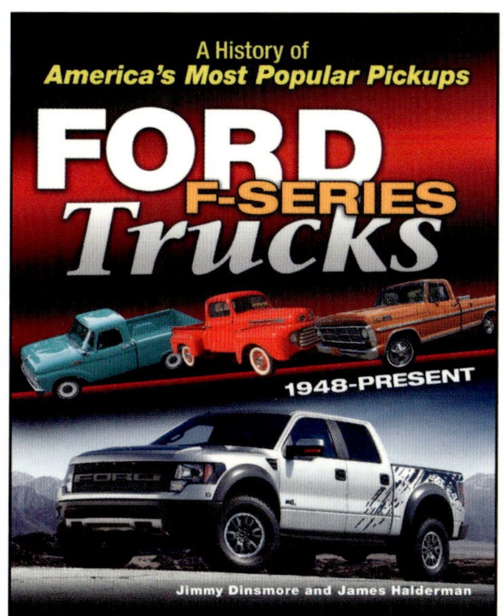


**FORD F-SERIES TRUCKS: 1948–Present**
*by Jimmy Dinsmore & James Halderman*
In *Ford F-Series Trucks: 1948–Present*, authors Jimmy Dinsmore and James Halderman thoroughly dissect the history of Ford F-Series pickup trucks as seen from a technical viewpoint. 8.5 x 11", 192 pgs, 400 photos, Sftbd. ISBN 9781613255124 ***Part # CT661***

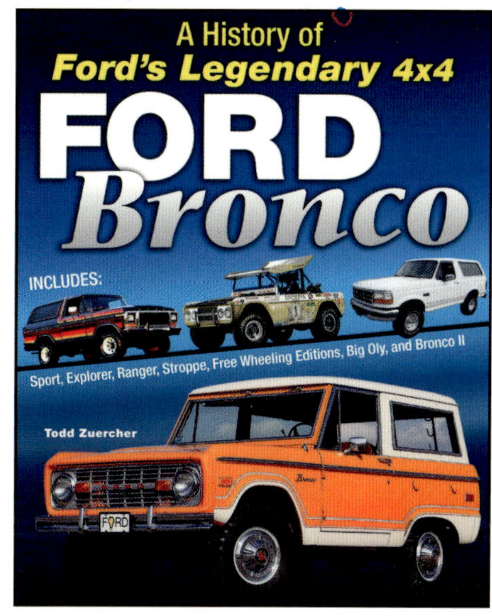


**FORD BRONCO: A HISTORY OF FORD'S LEGENDARY 4X4** *by Todd Zuercher* Technical details, rarely seen photos, and highlights of significant Bronco models along with the stories of those people whose lives have been intertwined with the Bronco for many years. This book will have new information for everyone and will be a must-have for longtime enthusiasts and new owners alike! Hardback, 8.5 x 11 inches, 192 pages, 90 b/w and 384 color photos. ***Item # CT634***

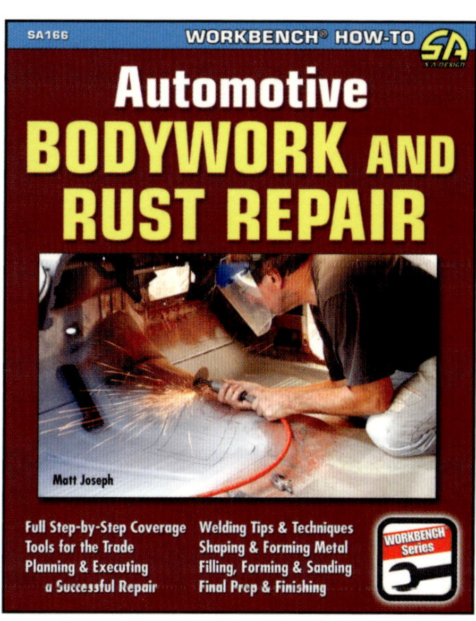


**AUTOMOTIVE BODYWORK AND RUST REPAIR** *by Matt Joseph* Author Matt Joseph shows you the ins and outs of tackling both simple and difficult rust and metalwork projects. He shows you how to select the proper tools for the job, common sense approaches to the task ahead of you, preparing and cleaning sheet metal, section fabrications and repair patches, welding options such as gas and electric, forming fitting and smoothing, cutting metal, final metal finishing including filling and sanding, the secrets of lead filling, making panels fit properly, and more. Softbound, 8.5 x 11 inches, 160 pages, 400 color photos. ***Item # SA166***

**www.cartechbooks.com or 1-800-551-4754**

# How To Rebuild FORD Power Stroke DIESEL Engines

Diesel-powered trucks have never bee more popular, and Ford's Power Stroke famil of Diesel engines have been a mainstay of th marketplace since introduced in 1994. Whi these engines are well known for their durabilit and longevity, they still require a rebuild a some point. How to best diagnose, repair, an rebuild these engines has been somewhat of mystery for those without access to the facto manuals—until now.

This book covers the vast majority of Pow Stroke Diesel engines on the road, and includ the story of their design. Each major system use within the Power Stroke engines is describe and discussed in detail, with full-color phot of every critical component. A complete ste by-step engine rebuild of both the 7.3L and 6.0 engines is also included.

Also included are a wide range of engir building tips, charts, graphs, and informatio packed sidebars to share even more informatio Detailed graphics show the engine's vario system designs.

Author Bob McDonald is a well-known Diesel engine expert. He is the owner of a Diesel repair specialty shop and has more than 20 years experience repairing and rebuilding both gas and Diesel engines of all kinds.

9 781934 709610 90000

**For a free catalog of all our books, write or call:**

**CarTech®**
6118 Main Street
North Branch, MN 55056
(651) 277-1200 / (800) 551-4754
www.cartechbooks.com

ISBN 978-1-934709-61-0
Item SA213
Written, edited, and designed in the U.S.
Printed in China

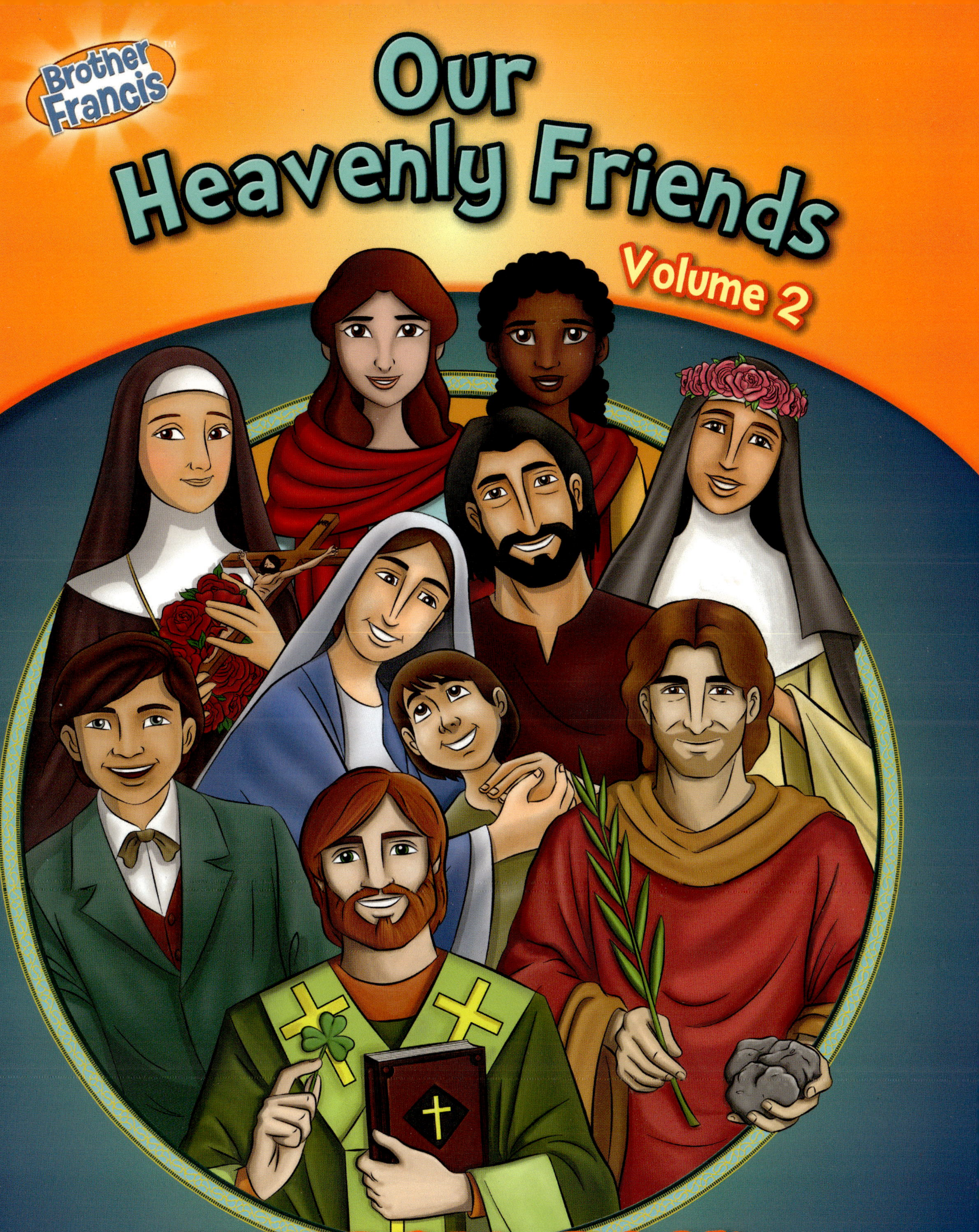
Brother Francis™
Our Heavenly Friends
Volume 2
COLORING BOOK

# My Heavenly Friends Prayer Card Collection

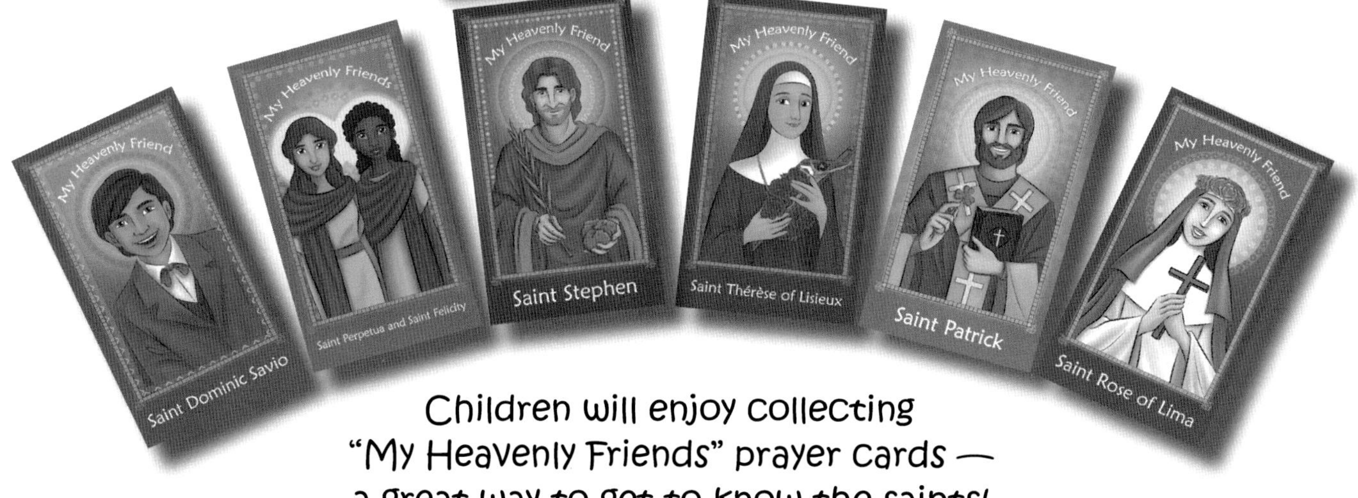


Children will enjoy collecting
"My Heavenly Friends" prayer cards —
a great way to get to know the saints!
Designed in **full color**, each laminated card
displays a captivating image and prayer,
and measures 2.5 x 4.5 inches.



Illustrations by: Herald Entertainment, Inc.
Visit our Brother Francis series website at: www.brotherfrancis.com
Printed in the United States of America
First Printing: February 2013
ISBN: 978-1-939182-15-9